Geography
for All

Judy Sebba

David Fulton Publishers

London

David Fulton Publishers Ltd
2 Barbon Close, London WC1N 3JX

First published in Great Britain by
David Fulton Publishers 1995

Note: The right of Judy Sebba to be identified as the author of this work has been asserted by her in accordance with the Copyright, Designs and Patents Act 1988.

Copyright © Judy Sebba

British Library Cataloguing in Publication Data

A catalogue record for this book is available from the British Library

ISBN 1-85346-307-8

Designed by Almac Ltd., London
Typeset by Harrington & Co
Printed in Great Britain by Bell & Bain Ltd., Glasgow

Contents

Acknowledgements

This book was written in response to the many teachers, assistants and advisers who have sought information and ideas about teaching geography to pupils with learning difficulties since this subject became a curricular entitlement for all pupils. It draws on the practices of many practitioners too numerous to list here by name but who have demonstrated creativity and commitment. Those who contributed specific examples are named alongside these in the text. Thanks to all of them for their willingness to share ideas and activities with a wider audience.

Special thanks go to those colleagues who have guided my own learning about teaching geography over the past few years, including Colin Conner, Martin Oldfield and the local authority advisory teams all over the country who have entered into debates about this area of work; to my AppleMac adviser, Mavis Robinson, always willing to provide further instructions when needed; to my children, Ian and Alan, for evaluating the geography software so willingly and tolerating a mother who rarely conforms to the stereotype.

Finally, it was in the early 1980s in a secondary school in Rochdale that this work really began, when my partner, John Clarke decided that he would teach a pupil with moderate learning difficulties the concept of contours by building three dimensional maps out of ceiling tiles, on our kitchen table. Hence, the debate about making geography accessible to *all* pupils began and continues. Thanks to John for the continued challenges, support and corrections to my geography and English.

Introduction

Aims of the Book

Geography is about the relationships between people and places. The subject itself is more accessible to pupils with learning difficulties than some other National Curriculum subjects. However, curricular development in geography has been hampered by three factors. First, there appears to be a lack of agreement about the content or the best way to teach it. Secondly, like many other subjects, it is often taught by those with little or no training in the subject. Lack of confidence in the subject may lead to a lack of clarity about learning outcomes. Finally, it is generally agreed that the National Curriculum requirements for geography, as specified in 1991, were poorly structured, overloaded and lacked coherence. Hence, in England this influence on curricular development in geography has often been unhelpful.

Curricular entitlement for all pupils has been welcomed by those working with pupils with learning difficulties. However, legal entitlement does little to ensure that meaningful access to a subject occurs in practice. For staff to develop appropriate activities and long-term plans which provide coherence, progression and continuity, a confident grasp of the subject will be needed. This requires staff development through courses, school-based activities and resources. This book aims to provide one resource to assist and encourage staff to develop an interest in and commitment to geography. Hence, many examples, ideas and activities are included in this book which have come from teachers working with pupils who have diverse needs in a range of settings. Often, effective practice and innovatory curricular development pass without notice in schools. Here, this practice is recognised and described for the benefit of others who are likely to be more inspired by work done by colleagues than by the theories of teacher educators.

An important aim of this book is to dispel myths about geography and to inspire confidence. For example, the views of school staff about geography may stem from their own experiences at school some (or many!) years ago. Hence, the subject is seen as being about capes and bays, rivers in little known countries and coffee production in South America. As this book will demonstrate, these elements are still part of the geography curriculum, but perhaps a smaller proportion of it, particularly at the earliest levels, than was previously the case. Geography has become recognised as a separate subject at primary level only since the introduction of the National Curriculum. Prior to this, and to the detriment of good teaching according to inspectors (DES, 1989), it was subsumed and sometimes lost within topic work.

The central aim of the book is to make geography more accessible to *all* pupils. Chapter 4 explicitly addresses the issue of access and considers ways of extending access through information technology and group work. The other chapters address access indirectly through providing a description of the essential skills, concepts and knowledge in geography, examining models of planning, recording and assessment and providing examples of practical activities and resources. It is intended that by familiarising staff with the subject and illustrating the work of others, readers will be inspired to further develop their own work.

Who Is It Aimed At?

There are two groups of staff who might find this book useful. The first are teachers who are working with pupils with learning difficulties and who have no background in geography. They will be mainly, but not exclusively, in primary and special schools. At the secondary school level, a significant minority of those teaching geography are non-specialists (OFSTED, 1993a). It is unrealistic to expect non-specialists effectively to meet the demands of teaching geography at this level. One consequence of this may be to give the non-specialist the lower ability classes in schools in which streaming or setting is used. This practice has been noted to lead to less effective teaching and learning (OFSTED, 1993a) since arguably these pupils need the most effective teachers in order to make progress. For teachers with little or no qualifications or experience of geography this book provides an introduction to the essential knowledge, skills and understanding at the earliest levels of geography work, illustrated by activities found to be helpful by other teachers.

The expertise of teachers who are qualified in the subject is not always used in primary schools (OFSTED, 1993b) or special schools. These teachers and the recognised specialists may be looking for examples of how the principles and practice in geography can be applied at the earliest levels, enabling the greatest possible access for pupils with learning difficulties. Teachers who have used methods and resources effectively with these pupils offer their ideas in this book. None claim to provide the perfect answers. Rather, they are shared as ongoing developments in the interest of professional collaboration.

People other than teachers, such as therapists, assistants, inspectors, advisers, lecturers and parents, may find items of interest in this book. It is important, for example, for therapists and parents to see the opportunities for communication or physical development that occur within geography and for ways in which activities outside the school day can be used to support learning.

Most importantly, this book is for the benefit of pupils with learning difficulties. If it contributes to the development of a broad, balanced and relevant curriculum for them which includes geography, I will have succeeded in my task.

Geography and the National Curriculum

In chapter 1 the skills and content in geography are described and links with other subjects are highlighted. There is no attempt in this book to relate the activities described specifically to content in the current legal requirements for geography in the National Curriculum. Most of the activities could be related to these, but in the interests of accepting the need for the precise content of the

National Curriculum to be regularly reviewed, the content and skills referred to are mostly those assumed to be central to geography, however it is defined. Other published resources mentioned do tend to refer to the current requirements and teachers will need to be mindful of which version is being used.

In the section on planning and recording, the examples cover the 1991 requirements but have been selected for consistency with the 1994 consultation document. No reference to attainment targets is made since their future use is likely to be limited to reporting. Instead, the activities focus on the geographical content or skills covered and ways of increasing access for all pupils.

Terminology

The ideas and examples in this book range from those demanding some reading and writing skills to those which demand no literacy skills and only limited or no communication skills. Some teachers in special schools are working with pupils who have extremely diverse needs. A few teachers in mainstream schools which have adopted 'inclusive education' policies or are linked to special schools teach similarly diverse groups. Hence, it is inappropriate to identify specific disability groups and link each to particular activities described. Throughout this book the term 'learning difficulties' is used since, while recognising the shortcomings of this term, it is currently most commonly used in schools. The exceptions to this principle are the references made to pupils with visual impairment, for whom geography presents specific accessing issues. These pupils are referred to in the context of particular activities that include observational skills or resources, the use of which demand good vision.

Where possible, gender stereotyping has been avoided by pluralising the pronouns. If specific pupils or teachers are being described, their actual gender is used, but elsewhere pupils are referred to as 'she' and teachers as 'he'. Pupils are consistently referred to as 'pupils with learning difficulties' in keeping with the 'people first' principle.

CHAPTER 1

What is Geography?

Introduction

Geography is about the relationships between people and places. It aims to help pupils make sense of their surroundings and develop an understanding about the interaction of people with the environment. In a very comprehensive review of geography education, Gregg and Leinhardt (1994) describe the discipline of geography as being characterised by four concerns. The first is the distribution of features over the earth's surface which contribute to the unique character of places (for example, mountains, rivers, sea, etc). The second is trying to understand how and why certain things happen where and when they do so (for example, volcanoes). The third is the ways in which things that occur are causes and consequences of human decision-making (for example, destruction of the rainforests). Finally, geography is concerned with communicating information and ideas through the language of maps. These four concerns interact with one another in a variety of ways rather than being mutually exclusive.

It is the interaction between the different aspects of geography which makes it difficult to split up into neat categories for descriptive purposes. The division which occurred in the 1991 National Curriculum requirements between geographical skills, places, physical, human and environmental geography resulted in a tendency to address one or two areas at a time in geography teaching thereby preventing pupils from understanding the relationships between the different elements (OFSTED, 1993b). Hence, the revision attempted to re-integrate the various aspects by encouraging teachers to address geographical skills and themes through the places on which they are focusing and by combining them where possible. However, in order to provide a logical structure for the description of activities in this book, these five aspects of geography will be used to classify the relevant skills, knowledge and understanding.

Some important questions (adapted from Wiegand, 1993, p.77) which geographers address include:

- What is this place like?
- Where is this place in relation to where I live or other places I know?
- How is it similar to, or different from, other places?
- How is this place changing?
- What would it be like to be in this place?
- What do I like about this place?
- How is this place connected with other places?

These form the basis for geographical enquiry which is addressed more fully later in this chapter.

Why Teach Geography?

The HMI Curriculum Matters document *Geography from 5 to 16* (DES, 1986) defined a different set of objectives for each of early primary, later primary and secondary work. The ones reproduced here are those from the early primary section since the others merely express more sophisticated versions of them, specifying the use of globes, maps and atlases and political and economic perspectives within a broader definition of place which includes an international dimension. The document suggests that geographical learning experiences should enable pupils to:

- extend their awareness of, and develop their interest in, their surroundings;
- observe accurately and develop simple skills of enquiry;
- identify and explore features of the local environment;
- distinguish between the variety of ways in which land is used and the variety of purposes for which buildings are constructed;
- recognise and investigate changes taking place in the local area;
- relate different types of human activity to specific places within the area;
- develop concepts which enable them to recognise the relative position and spatial attributes of features within the environment;
- understand some of the ways in which the local environment affects people's lives;
- develop an awareness of seasonal changes of weather and of the effects which weather conditions have on the growth of plants, on the lives of animals and on their own and other people's activities;
- gain some understanding of the different contributions which a variety of individuals and services make to the life of the local community;
- begin to develop an interest in people and places beyond their immediate experience;
- develop an awareness of cultural and ethnic diversity within our society, while recognising the similarity of activities, interests and aspirations of different people;
- extend and refine their vocabulary and develop language skills;
- develop mathematical concepts and number skills;
- develop their competence to communicate in a variety of forms, including pictures, drawings, simple diagrams and maps. (DES, 1986, pp 5-6)

The contribution which geography makes to learning in other curricular areas, reflected in the last three of these objectives is considered in more detail later in this chapter.

Requirements of the Geography Curriculum

The requirements of the geography curriculum include five areas of geographical work:

- geographical skills
- places
- physical geography
- human geography
- environmental geography

These five areas correspond to the attainment targets in the 1991 geography National Curriculum but are incorporated into the programmes of study in the proposed revision for 1995. This partly reflects the changing structure of the National Curriculum in which attainment targets have been substantially reduced and are to be used as a basis for reporting to parents, not for planning, teaching or assessment. It also reflects the view that the inter-relationships between these areas makes their separation unhelpful. For the purposes of providing a basic introduction to the requirements, these distinctions will be retained. Whatever future revisions to the curriculum take place, these aspects are seen as fundamental to the subject, they are reflected in the objectives listed above and will continue to be included in some form.

Geographical skills

Geographical skills encompass much of what geography is often perceived to be about – maps and field work techniques. They are essentially about the skills needed to study geography, some of which – for example measuring – are also required in other areas of the curriculum. The skills include following directions, using plans, maps, globes, atlases, grids and coordinates, and symbols and keys. Pupils are required to make maps as well as read them.

Field work techniques involve use of compasses, measurement, for example of aspects of the weather, and relevant use of information technology such as databases and spreadsheets. Field work need not involve developing these skills in specific places renowned for their geographical features. The earliest development of the skills may occur in the immediate environment.

Places

The study of places has been one of the more controversial areas of geography. The debate centres around two issues. One is similar to the issue concerning the balance of facts to skills which should be taught in history. The amount of factual knowledge that should be required in geography remains a contentious issue. There are those who consider that there is too much emphasis on skills while others wish to protect or extend skills work. Wiegand (1993) argues that the under-emphasis on enquiry in the curricular requirements could be due to concerns that enquiry approaches encourage children to question the world around them and that politicians have a vested interest in the uncritical transmission of facts.

However, as in history, skills cannot be taught in isolation. Geographical enquiry focuses on an area, however broadly defined, ensuring that opportunities for learning skills and facts are offered simultaneously. One of the concerns about the 1991 National Curriculum requirements was that there was an overload of precise pieces of place knowledge – for example, requiring pupils to identify and name the seven continents, oceans, tropics, numerous countries, cities, rivers and mountainous regions from maps at a very early stage (see Clarke, 1992 for more detailed comments on this). Many of these demands have now been relocated into the requirements at more advanced stages.

The second issue is about the stage at which less familiar places should be studied and the balance between local, national and international studies. In recent years this has been complicated by the need to ensure more emphasis on European studies. It is clear that for very young children or pupils with substantial learning difficulties it is helpful to develop basic geographical skills in a familiar context. However, some teachers fear that failing to encourage study of areas beyond the home region early enough will encourage parochialism. In 1989, the report on teaching and learning in history and geography (DES, 1989) noted that in primary schools there was almost total absence of a national and world dimension to the geography curriculum. The OFSTED (1993b) report further substantiated this concern by suggesting that too little emphasis was put on the study of places beyond the local area. Furthermore, Wiegand (1993) suggests that Britain and Europe are over-emphasised in the place requirements at the expense of the wider world. Some pupils develop an interest in distant places through coverage about them on the television. This debate remains unresolved. For all pupils it is necessary to balance relevance and meaningful learning with the need to develop an interest in issues and cultures beyond the immediate vicinity.

Study of places includes identification and naming places on maps, atlases and globes and through observation in the context of field work. It involves identifying distinctive features of places and recognising similarities and differences between places. Local studies are included in the study of places at the earlier levels and are defined in terms of an increasing spatial area as the pupil progresses. Hence, at the earliest stages the school building and its immediate locality is studied and this progresses to an area around the school beyond the immediate vicinity. The relationship between themes and issues in particular locations is also important. For example, studying the reasons for the location of an industry in a particular area might provide an insight into the relationship between availability of raw material, terrain and costs.

Physical geography

Physical geography covers the study of the weather and climate including the seasons. It also covers rivers, land forms, vegetation and soil. While overlap between physical geography and science requirements has been reduced, it is still evident in the requirements. For pupils working at the earliest levels, the human aspects of physical geography, such as the implications of the weather, may be more accessible than studying features of physical geography in isolation.

Human geography

Human geography covers settlement, population, communication and movement, including transport and economic aspects. According to the OFSTED report (1993b) on the teaching of geography, human geography has been under emphasised compared with the attention given to maps and weather. This is surprising given the accessibility of aspects of human geography such as journeys, jobs, transport and homes.

Environmental geography

Environmental geography is about the use and misuse of natural resources, the quality of different environments and protecting and managing the environment. This has been a contentious area of geography as it has been seen as providing a potential political platform for environmentalists. In spite of this concern, it has survived the revision and continues to provide tremendous scope for studies that are relevant and meaningful, sometimes in the local area. Interestingly, this area of geography was another seen as not receiving enough emphasis (OFSTED, 1993b) as compared to aspects of physical geography and geographical skills.

Geographical enquiry

Geographical enquiry is seen as integral to the content as well as the teaching methodology of the subject. Hence, an enquiry approach is specified in the requirements in geography whereas prescribing teaching methodology is avoided in general, in this and other subjects. Enquiry encourages pupils at the simplest level to ask questions and search for answers, and at a more sophisticated level to develop and test hypotheses in order to enable geographical concepts and content to be better understood. The National

---

Asking questions

Why do people send postcards?
Have you received some?

Planning

What can we find out from a postcard?
Pupils define a few questions for the investigation of their own postcards.

Investigating

What can you see in the picture?
What is happening in the picture?
Where is the postcard from?
Pupils discuss the picture and agree answers to the questions. They word-process list of features, draw and label them and search in atlases.

Evaluating

Would you like to visit this place?
What could it be like to live there?
How could you find out more about this place?
Pupils discuss and agree answers to questions. They record their answers.
They study the stamps and postmarks and copy the key words.

Figure 1   Geographical enquiry: an example (adapted from National Curriculum Council 1993a, p.28, with the permission of the School Curriculum and Assessment Authority)

Curriculum Council (1993a) INSET resources describe the route of enquiry as asking questions based on existing knowledge, planning an enquiry, carrying out the investigation and evaluating the conclusions. An example is given in this publication in which post cards from localities beyond the UK received by the pupils are the focus of an enquiry (see figure 1).

At the heart of the enquiry process is effective questioning which can be used to promote curiosity and interest and ensure active involvement in the learning process (Curriculum Council for Wales, 1991). Progression in enquiry is planned through increasing the levels of exploration and analysis required and through the use of more complex and wide-ranging sources of information.

Cross-curricular Skills

Geography promotes the cross-curricular skills of communication, information technology, numeracy, personal and social skills, problem-solving and study skills. Geographical enquiries involve all these skills and can provide an interesting and relevant context in which they develop. For example, pupils might conduct a survey of the local shops used by the members of their family as illustrated in figure 22 in chapter 3. This could involve interviewing, data handling (perhaps using the computer to display information), trying to provide reasons why certain shops are used more frequently and reporting their findings. There is considerable overlap between these skills and the criteria for quality of learning used on school inspections (OFSTED, 1994).

Cross-curricular Themes

The links between the cross-curricular themes such as environmental education, careers education and economic awareness and geography are extensive. Staff working with pupils with learning difficulties welcomed the cross-curricular themes as a vehicle for relating work in the National Curriculum subjects to the life skills activities that were well established in their schools. Activities designed to address the themes often provide excellent opportunities to promote the enquiry skills needed to study geography such as observation, recording, data handling and interpretation. The themes are fully described in the *Curriculum Guidance* series (National Curriculum Council, 1990a, b, c, d, e).

Economic and industrial understanding

Geography contributes to economic and industrial understanding for all pupils through activities on human and environmental geography. Activities which address work and leisure, goods and services, transport and settlement all contribute to both human geography and economic and industrial understanding. These can be taught at the most basic level, for example, by looking at different types of transport. At a more sophisticated level, issues relating to the use of natural resources, the impact of industry on the environment and waste disposal contribute to this theme and to environmental geography. In chapter 4, activities are described involving simulations of planning and environmental issues which contribute directly to both geography and economic understanding.

Health education

Health and safety issues overlap largely, but not exclusively, with environmental geography. Field work offers an ideal context for developing safety awareness. The quality of the environment, pollution, conservation and the health of communities all contribute to both geography and health education. Many of these issues are also relevant to environmental education.

Environmental education

Environmental education contributes to geography in its entirety. The National Curriculum Council (1990d) describes the theme as including climate, soil, rocks and minerals, water, energy, plants and animals, people and communities, and buildings, industrialisation and waste. Hence, environmental education addresses physical, human and environmental geography. Environmental education can be promoted at all levels including pupils experiencing profound and multiple learning difficulties, as the example in figure 2 illustrates.

Each group of children, blindfolded and led by adults (teachers and parents) walked along the trail, holding a rope.

Various stops were selected to encourage them to use their senses in the discovery of the natural world:

● touching the texture of a wall, shapes and sizes of the bricks

● smelling the herb garden, the rosemary and bluebells

● walking (travelling in a wheelchair) across the gravel and listening to the sound of their feet (the wheels)

● listening to the sounds in an open area

● touching the trees, feeling the barks

After they had uncovered their eyes, some children were asked to find their way back to the centre, using the various clues they had picked up during the trail.

The activity encourages the pupils to explore the environment and some pupils may be able to appreciate the need to protect and improve the environment. For pupils for whom it is appropriate, the activity could be extended to drawing the trail on a plan of the site and adding their own 'observations'.

Figure 2   The Nature Trail (adapted from National Curriculum Council 1990d, pp.20-21, with the permission of the School Curriculum and Assessment Authority)

Careers education and guidance

Careers education provides an appropriate context for considering the different types of work which people do and the reasons for variations in this. Hence, it can make a significant contribution to human geography and may venture into environmental geography through consideration of environmental issues in industry. Many pupils have the opportunity to undertake work experience and this provides an opportunity to explore these issues. In chapter 3, an activity is described in which pupils with learning difficulties undertake work experience in their own school as a basis for exploring their understanding of adult roles.

Citizenship

The issue of human rights is fundamental to citizenship and may arise in various areas of geographical work, for example, within planning and environmental issues and distribution of world resources. The relationship between values and beliefs and decision-making is also central to this theme although it demands more sophisticated understanding than basic human rights issues. Citizenship also involves looking at roles and responsibilities, communities, work, employment and leisure, and public services. This theme may be particularly pertinent for pupils with emotional and behavioural difficulties for whom addressing responsibilities, relationships and value systems are major priorities.

Cross-curricular Dimensions

Geography can promote equal opportunities and contribute to multicultural understanding by challenging stereotypes and the use of sexist language. Human geography offers the opportunity to examine roles and responsibilities, job opportunities and economic factors. In the survey of history and geography in primary schools (DES, 1989), it was noted that only in a small minority of the schools was cultural diversity being promoted through the teaching of geography. Cultural diversity can be presented positively through the choice of materials and teaching techniques used to study different societies. In the *Geography Non-Statutory Guidance* (National Curriculum Council, 1991), it is suggested that when comparing societies or groups of people, considering the similarities first and differences second can help to promote positive images and challenge myths and misconceptions.

Spiritual, Moral, Social and Cultural Development

Pupils' spiritual, moral, social and cultural development is one of the major focuses of the OFSTED (1994) inspection model. However, there is nothing new in the idea that schools and teachers are responsible for promoting development in these areas. In fact there is a strong argument that this is one of the main reasons why schools exist. A few examples of the contributions which geography can make are given in the guidance section of the *Handbook for the Inspection of Schools* (OFSTED, 1994) but the importance of the subject in these matters is insufficiently recognised there. Much of the study of geography is concerned with the consideration of values and, as such, the subject has a major, not a passing, contribution to make.

In the first place, much of geography deals with people – with their beliefs and values and the effects these have on their behaviour. It deals with the human condition – with the search for survival, prosperity and happiness often against many odds, both human and natural. The study of other people, both those who are near to home and those further afield, offers opportunities for the teacher to get pupils to reflect on aspects of their own lives through comparison and contrast and thus foster their spiritual development.

Much of human and environmental geography deals with moral issues and, perhaps pre-eminently, allows pupils to see that telling right from wrong is not always easy. 'Issue based' geography does this well. A simple by-pass study allows pupils to see that it is not necessarily 'right' for the local publican or village shop to have the village by-passed although it might be 'right' for the children who have to cross the road on their way to school. Geography has an important contribution to make to the understanding of moral principles and often gives insight into the difficulties and contradictions which can arise in seeking to respect other people, justice and property.

There are 'personal' aspects of pupils' social development. These are fostered by the way the school works and the way people treat one another. They are also fostered by the opportunities which are offered to work collaboratively which stem from the teaching methods employed. But social development is also about the ways in which societies are organised and geography has contributions to make here. The *Handbook for the Inspection of Schools* specifically states that social development is to be judged by the extent to which pupils gain an understanding of how societies function and are organised in structures such as the family, the school and local and wider communities. Where else but in geography and RE will pupils study wider communities?

Schools are also about preparing pupils to understand aspects of their own culture, be these religious, social, aesthetic or ethnic. In its study of other societies, geography has an important role to play. This is particularly so in the work which geography should be doing in the challenging of stereotypes, through the value which the subject should be giving to other people in getting beneath the skin of their cultures and looking for reason and explanation. As with history, in geography the study of other cultures is a good way of better understanding your own.

Links Between Geography and Other Subjects

The links between history and geography have been assumed since the subjects were often taught together as part of a topic or integrated humanities course. However, the strong links that each of these subjects has with other subjects – for example, history with English and geography with mathematics – have been relatively unacknowledged. In the survey of history and geography in primary schools (DES, 1989), it was noted that the scientific and mathematical potential of geographical skills was only occasionally exploited.

Occasionally, a pupil may demonstrate an understanding of a concept within one subject when he or she could not do so within another subject. For example, in one activity described in chapter 3 involving grid boards to teach coordinates, a pupil demonstrated greater understanding during the grid game in geography than he had during a mathematics session. This may reflect differences in teaching or learning styles but will require teachers to swallow their pride in order to collaborate. This is further considered in the next chapter in the section on the role of the subject coordinator.

Science

Geographical enquiry involves similar skills to those required for scientific, mathematical, technological or historical enquiry, for example, hypothesising, observing, recording and interpreting. Traditionally, the greatest area of overlap between geography and science was the weather. In the proposed revision of the National Curriculum (for 1995) the weather has been removed from Science but remains in Geography. Links which remain include the water cycle, weathering of soil and rocks, tectonic processes and managing the environment.

English

Work in all subjects requires pupils to communicate and geography is no exception. In chapter 3, examples are given of teaching geography through stories. Geographical enquiry may involve research skills such as interviewing and use of a library, taking notes, dictating or using symbols or computers to record information, discussion with pupils and teachers and production of reports in whatever form is appropriate. Sometimes pupils may be expected to role-play as part of a simulation (see chapter 4) or work in groups on an enquiry. Geography also provides the opportunity to make or read different types of material such as labels, posters, signs, plans, maps, globes and atlases.

Mathematics

There are extensive links between geography and mathematics through numeracy, measurement, distance, direction, area, density, size, shape, scale, data handling and coordinates. The Curriculum Council for Wales (1991) note the importance of using the same terminology and computational techniques in the two subjects to maximise pupils' understanding. Practical work in geography, such as field work, construction of relief models and maps, provides opportunities for promoting mathematical concepts.

Technology

The use of information technology to increase access in geography and further develop information technology skills is addressed in chapter 4. Design technology and geography have links through studying technological development in different parts of the country or world, considering how problems such as water supplies are solved or remain unresolved in different parts of the world. Practical activities in geography provide an opportunity to promote technological skills by enabling pupils to decide what is needed – for example, to measure rainfall – how it could be made, what materials are required, and to evaluate the finished product.

History

The links between history and geography tend to have been better exploited since, traditionally, they have been combined for planning purposes. The links are particularly strong within the local study in which the opportunity is often taken to study the local area and how it has developed over time. Concepts such as settlement, industrial and economic development, soil erosion and rock formation require study over a period of time. Practically based topics for young pupils or older pupils with learning difficulties, such as houses and homes, people who help us, food and farming, our school and ourselves, all lend themselves well to both history and geography. The next chapter considers the planning implications of maintaining this approach.

Modern foreign languages

This subject has traditionally been assumed to be least accessible to pupils with learning difficulties. However, the National Curriculum requirements have been, and remain, sensitive to the whole range of pupils' needs. Included in the requirements is developing cultural awareness which links in with geography through the consideration of similarities and differences between their own culture and that of the country or community where the target language is spoken. In addition, the areas of experience include the world around us (travel, transport, etc.), the world of work, communications and the international world (tourism, global issues, etc.). Some schools have developed links with a school in France or Germany as a resource for work in modern languages and geography.

Music

Music is one aspect of culture and therefore provides one focus for studies of other countries or communities. Within the context of studying a different region or country it is appropriate to consider the similarities and differences between the music from that area and that from the home area.

Art

Art is another aspect of culture which can provide a focus for studies of other communities or countries. Within local studies it is likely that the architecture of local buildings will be considered. The materials used to construct buildings may also be discussed in the context of geography and history. When constructing maps or models in geography, pupils should be encouraged to experiment with a range of materials, tools and techniques in order to create the desired effects. Construction of maps or trails for others to follow will involve reviewing what happened and whether the accuracy, scale or proportions were appropriate.

Physical education

Links between geography and physical education are clear in the area of outdoor pursuits such as nature trails or orienteering. Within the context of dance pupils are expected to explore a variety of dance from different cultures. Within the context of gymnastics we have worked on directional skills by creating mazes using 'soft play' or other apparatus. The use of physical education activities as a context for geography is further described in chapter 3.

There is little doubt that the 1991 curricular requirements in geography were unworkable for many teachers and pupils due to the overload in the content specified. Furthermore, as Clarke (1992) has made clear, the structure of the requirements lacked coherence and led to the necessity to teach different pupils about different parts of the world simultaneously in order to meet their diverse needs. While this situation has not been entirely overcome, the drastic reduction in prescribed content and more coherent structure proposed enables some level of planning for diversity. The previously prescribed list of pieces of place knowledge at the earliest levels made parts of the geography requirements impossible to meet in the time allocated. For pupils with learning difficulties this emphasis on facts was unhelpful.

The extensive content led teachers away from the more interesting styles of teaching that some had adopted in geography including simulations, role play, and problem solving in groups. The need to cover this content, meet diverse needs and provide different content for different pupils led, in some cases, back to dependence on work sheets, a practice least likely to benefit the pupil with learning difficulties. The use of more inspiring techniques is helpful to pupils with learning difficulties and if a reduction in prescribed content leads to moves back to more extensive use of these techniques, pupils with learning difficulties may particularly benefit.

Four particular implications of the curricular requirements for pupils with learning difficulties relate to wider organisational factors. These are breadth and balance, assessment, subject specialist teaching and resources. Breadth and balance in the curriculum may be difficult to maintain when the teaching time available is reduced by other priorities. For example, teachers working with pupils with emotional and behavioural difficulties may need to plan across a key stage selecting an appropriate amount of content to cover realistically in the time available, given likely interruptions, need for individual counselling and group discussion focusing on behaviour.

The assessment issues arise from the hierarchical nature of the National Curriculum which demanded that in every subject, skills were defined through statements of attainment in the order in which pupils learn them. It is not always clear that we can identify, with confidence, the order in which pupils acquire geographical skills. Pupils with learning difficulties may remain on the earliest levels throughout their school career so relatively arbitrary ordering of skills may worsen this problem. However, the restructuring of the National Curriculum has partly overcome this problem since the focus will be more clearly on the programmes of study. Any approach to assessment must ensure that it is geographical skills and understanding that are being assessed and not literacy and communication.

The lack of subject specialist teachers in primary and special schools is a potential problem for all pupils. However, pupils with learning difficulties may require a concept to be presented through several different approaches before it is fully understood. The subject specialist is more likely to have the confidence in the subject to do this. More teachers are being given the opportunity to undertake subject training in geography, which appears to develop the confidence needed to plan more selectively while maintaining breadth and balance. However, in both primary and special schools, many teachers are required to coordinate two or more subjects which limits the extent to which any one specialism can be developed.

Traditionally, resources in geography teaching have been more extensive in the secondary phase. The introduction of the National Curriculum has led to many publishers producing materials on primary geography, some of which have been tied in specifically with the current requirements and are therefore unlikely to withstand the test of time given regular revisions. The published material available that specifically relates to pupils with learning difficulties is still very limited and includes examples produced by groups of teachers working with their LEA advisory services (Devon County Council, 1993; East Sussex County Council 1991; Humberside County Council, 1992). These are all very useful as they describe activities tried, tested and evaluated by teachers. Examples from each of them are given in chapter 3.

This chapter has provided a definition of geography, purposes of teaching geography, curricular requirements and links with cross-curricular themes and other subjects. Some implications of the curricular requirements for pupils with learning difficulties have been offered. The next chapter considers planning, recording and assessment in geography within a framework of accessing geography for all pupils.

CHAPTER 2

Planning, Recording and Assessment

Introduction

The main criticisms of previous geography teaching reflect issues relating to planning, recording and assessment. Poorly planned and coordinated topic work was seen as leading to superficial treatment of geography (OFSTED, 1993a). Insufficient long-term planning led to lack of breadth and balance within the key stages, and lack of cooperation between primary and secondary schools limited continuity and progression across key stages. Many of the deficiencies in planning were considered to be partly due to lack of subject expertise and the complexity of the 1991 National Curriculum requirements (OFSTED, 1993a). Recording which lacked rigour failed to identify the progress in geography made by individual pupils.

This chapter provides some examples of planning, recording and assessment used in teaching geography to pupils with learning difficulties. Other publications that are particularly helpful in this area include the *Non-Statutory Guidance* for geography (Curriculum Council for Wales, 1991; National Curriculum Council, 1991), the INSET Resources from the National Curriculum Council (1993a & b), and the key stage 2 planning document from the National Curriculum Council (1993c). Palmer (1994) provides many examples of good practice including formats for assessment and recording (by pupils and teachers) in geography. These resources include a useful range of frameworks for planning within which the content and specific examples developed will obviously need to relate to the needs of the particular pupils involved.

The planning of the geography curriculum in a school may involve development of a policy statement and will involve plans covering a whole key stage and across key stages, schemes of work and a format for assessment and recording reflecting the whole school policy for these. To assist clarity, these levels of planning have been categorised as long-term, medium-term and short-term in this chapter as follows:

- **Long-term planning** Planning across a key stage, two or three key stages, department or whole school. Indicates main topics/units to be covered and cycle to ensure breadth, continuity and progression.
- **Medium-term planning** Planning schemes of work which detail content,

methods and resources to be used in each unit/topic.

- **Short-term planning** Detailed activity or lesson plans. While these may be unnecessary for subject specialists for whom the schemes of work will provide sufficient guidance, the non-specialist may find it useful to plan activities in detail in the initial stages of development of teaching geography. These are also useful as a basis for evaluating the effectiveness of geography teaching in the school and for generating a bank of activities to assist other teachers where class-based teachers are required to teach many subjects.

Policy Statements

A policy statement could be a relatively short document providing a general overview of the school's policies on geography teaching. It should reflect the values and general aims that underlie geography teaching in the school. It will include:

- **Aims and objectives** These should cover the definition of geography, statements about breadth, balance, relevance and differentiation. In some schools, it may be particularly important to establish clearly how appropriate geographical activities will be distinguished from tokenistic references to geography in topic plans or recording sheets. The contribution geography can make to other curricular areas, cross-curricular themes and learning skills should be drawn out. A statement on the planning, recording and assessing of geography should be included. The policy statement can refer to the schemes of work in order to avoid overlap and repetition between the two.
- **General principles on methods** These might define the priorities in teaching geography, how it will be delivered and what resources will be used (all in outline). The particular issues relating to implications for pupils with learning difficulties should be mentioned. For example, methods of ensuring access for pupils with visual or literacy difficulties should be described.

The policy statement should reflect guidance and policies of the local authority (if the school is under LEA control) and national requirements. It must be agreed by the staff and governors and regularly monitored and reviewed. Where appropriate, it may be related to the school development plan and the staff development plan.

An example of a policy statement for geography from a special school is given in figure 3.

---

1. INTRODUCTION

1.1 The teaching of Geography at the Edith Borthwick School will be considered within this policy statement.

1.2 Geography is a National Curriculum foundation subject and is compulsory to the end of Key Stage 3.

1.3 At the Edith Borthwick School we feel it is essential that all children and young people experience Geography. Such provision, however, shall be at a level appropriate to their development and abilities.

1.4 The teaching of Geography at the Edith Borthwick School is based on the premise that the more remote the concepts are from pupils' own experience, the less relevant those concepts are to their needs.

1.5 A consequence of 1.3 and 1.4 above is the fact that students in both departments of the school may well, at Key Stage 3, be working on programmes of study at or below level 1.

2. THE AIMS OF TEACHING GEOGRAPHY

2.1 The main aim of Geography at the Edith Borthwick School is to promote an awareness of the world around them and the variety of life within it.

2.2 Specifically this may be done by encouraging within the pupil the development of:

a) an awareness and interest in self and the immediate surroundings;
b) identification and observation and enquiry skills through the exploration of first local then wider environments;
c) the ability to distinguish between variety of land use and the variety of purposes for which buildings are constructed;
d) ability to recognise and investigate change;
e) an ability to relate different types of human activity to specific places;
f) concepts and language which enable them to recognise and describe relative position and spatial distribution within their own environments;
g) an awareness of seasonal changes of weather and of effects which weather conditions have on the growth of plants, on the lives of animals and on their own and other people's activities;
h) an understanding of the different contributions which a variety of individuals and services make to a community;
i) an interest in and knowledge of places and peoples beyond their immediate experience;
j) an ability to use a variety of maps effectively;
k) a competence in communicating in a variety of forms including pictures, drawings, simple diagrams and maps.

2.3 Wherever possible the above involves practical use of the school buildings, grounds and the local village. A wide variety of visits to areas of interest outside the immediate locality are planned termly as part of the curriculum.

3. GEOGRAPHY DELIVERY

3.1 Geography is delivered through a topic based approach in Key Stages 1 and 2 and through a modular approach in Key Stage 3.

3.2 In addition to 3.1 much Geography is incidental in nature and does not form part of the formal curriculum planning process. Examples of 'incidental' Geography include:
- finding their way around the school
- running errands
- shopping in local village/using the post office
- recording/discussing the weather daily.

3.3 On leaving school the students will have experienced work covering the aims set out above where appropriate to their individual needs. The level of achievement will be recorded in the school's recording system.

4. PLANNING

4.1 All teachers prepare termly topic or module plans indicating which aspects of the National Curriculum and other areas of learning they intend to cover.

4.2 Teachers are encouraged to collaborate within their departments and across phases to ensure a level of cohesion within Geography provision.

5. REPORTING

5.1 Pupils' experiences in all areas and aspects of the curriculum are reported to parents on the annual progress report at the end of the summer term.

5.2 Parents have the opportunity to discuss the report with the child's teacher on Parents' Evening and to express their views on a Parents' Questionnaire.

6. INSET AND SUPPORT

6.1 A school based Geography team has been set up, comprising representatives from all areas of the school, to augment, advise on and support the development of Geography provision throughout the school.

6.2 The county's advisers are available for support and advice.

6.3 Those staff with responsibilities in Geography are funded for INSET under the school's Staff Development Plan.

Figure 3 A policy statement for geography from the Edith Borthwick School, Essex (developed by Jacqueline Hammacott, Geography Coordinator)

Long-term Planning

The long-term planning should ensure breadth and balance within each key stage and continuity and progression across key stages. It will assist in decisions regarding the allocation of time and can be used to identify potential links to other subjects. Figure 4 provides an example of a key stage 1 curriculum plan adapted from the non-statutory guidance (National Curriculum Council, 1991, C5). It illustrates the use of broad topics within which there is clear geography coverage planned and which provide breadth and balance in geography across the key stage.

The implication is that topics should be determined ahead on a three year cycle or according to the number of year groups in that key stage. Time allocations for geography will vary each term but should be appropriate across each year. In primary and special schools which have mixed year classes, careful planning will be needed to ensure breadth and balance and avoid repetition for each pupil.

Long-term plans should be subjected to scrutiny once developed. Some suggestions for appropriate questions that might be asked are given in figure 5.

	Autumn	Spring	Summer
Reception	**MY HOME:** Name their home area and country Why do people move homes?		**THE SEASONS** Weather in different parts of the world Similarities and differences with the local area
Year 1	**WHERE I LIVE** Features of the local area Land use Work and leisure activities Settlements	**GROWING THINGS** Rocks, soil and water Landscape Weather conditions	**PEOPLE WHO HELP US** Goods and services provided in the local area Work roles
Year 2	**HOW I GET TO SCHOOL** Why people make journeys Types of transport used	**SHOPS AND SHOPPING** Types of shops Locate on globe origins of goods Work roles Use of buildings Transport	**CONTRASTING LOCALITY** Features of a contrasting locality Effects on people's lives Similarities and differences with other localities Use of buildings Function, origin and size of settlements

Figure 4 An example of a key stage 1 curriculum plan (adapted from National Curriculum Council, 1991, C5, with permission of the School Curriculum and Assessment Authority)

- Are individual pupils' needs being met?
- Are the curricular requirements being met?
- Is there a progression across the years?
- How does geography relate to other curricular areas?
- Does the structure ensure continuity for individual pupils even when mixed year groups exist?
- Is there a balanced coverage of the areas, for example skills, places, physical, human and environmental geography?
- Does the plan allow for appropriate use of field work?
- Is a broad range of teaching approaches, for example, visits, role play and group work, implied?
- Are there adequate opportunities for geographical enquiry?

Figure 5 Monitoring and evaluating long-term plans

These questions will provide a suitable basis for an audit once the planning cycle has been implemented for a specified time (usually a year). One method of tracking curricular coverage in geography is described in Humberside County Council's (1992) guidelines on geography for pupils with learning difficulties. This audit should be part of the curricular audit that informs the development plan, enabling resource or staff development implications to be addressed.

Medium-term Planning

Medium-term planning provides the detail of the content to be covered across each key stage, specifying teaching methods and assessment opportunities. It should also indicate resources to be used and strategies for differentiation. Planning at this level is done through schemes of work which do not detail an individual lesson or activity but provide a description of the content to be covered in one subject or curricular area. Some documents refer to each scheme of work as corresponding to a subject with 'units' of work, one for each area of geography described within it. This introduces a fourth level of planning which might be confusing. This chapter therefore assumes that each area of geography, for example places, should have a corresponding scheme of work.

Scheme of work

A scheme of work should specify:

- aims and objectives: the knowledge, skills, concepts and attitudes to be promoted in this area;
- content: the area to be covered, related to the programmes of study in geography;
- methods: teaching approaches to be used, pupil groupings, etc.;
- differentiation: how the diverse needs of pupils in the class will be met;
- cross-curricular links/elements: links with other subjects, coverage of the cross-curricular skills, themes and dimensions;
- resources: published and unpublished materials, site visits, maps, video, photographs and other resources to be used;
- recording and assessment: how these will be done, opportunities within this area for assessment.
- monitoring and evaluation: how the subject coordinator/ head of department will ensure this scheme of work is reviewed and developed as necessary in response to review.

An example of an outline scheme of work from the National Curriculum Council (1991, p.C20) is provided in figure 6.

Unit of work for a KS2 class (year 6) – Study of another locality

Key Questions	Learning Objectives	Pupil Activities	Resources	Assessment Objectives
What are the similarities and differences between the local area and visited locality	**Concepts** Changing landscape Land-use The protection of the environment	**In the field:** group/individual observation, sketching Observation of flora, fauna and rocks Beach transect	Ordnance Survey 1:50,000 maps Large scale plans	Teacher assessment – oral by listening to comments and asking questions
How is tourism important to that locality	**Skills** Enquiry skills – data collection, observation, investigation Map work Atlas work Use of primary and secondary sources **Content** Wearing away of sea coast Depositing the eroded material Protection of coast Use of coastal areas for leisure activities, hotels, guest houses, retirement homes Vegetation	**Simple group work** to measure, record, sketch, photograph Use of compass Observation and sketching of a landscape Discussion of land-use Orientate OS map on site; on beach; at viewpoint; at monument, eg castle, keep Draw sketch maps Draw cliff face sketch and label Record temperature each day of visit to compare with year 5's log book at school, recorded same time	AA/RAC atlases, information books, tourist brochures/maps, aerial photos Teacher/pupil resources from locality to be visited Teacher's Centre in the locality Photos, postcards, spreadsheets, data programme Clinometers Compasses Cameras Hand lens Sample jars and bags	Pupil draws transect to scale Peer observation/ discussion Teacher observation Mount and label photos, comments on land-use, etc. Individual factual writing, drawing, folder production

Links with other subjects – mathematics, English, science.
Cross-curricular themes – environmental education.

Figure 6   An outline scheme of work (NCC, 1991, p.C20, reproduced with permission of the School Curriculum and Assessment Authority).

The need to involve pupils in planning, assessment, recording and reporting has been recognised in the handbook for inspection of schools (OFSTED, 1994) and Code of Practice (DfE, 1994) for implementing the 1993 Education Act. The curricular requirements for English, technology, art, music and physical education all refer explicitly to pupils reviewing and evaluating their own work. The criteria for quality of teaching used on school inspections makes specific reference to the need for teachers to provide opportunities for pupils to be made aware of the objectives of lessons. If pupils are more aware of why they are doing an activity and what they are expected to achieve, they will be better equipped to learn. Figure 7 from Beyton Middle School in Suffolk provides a clear illustration of how a plan of work may be shared with the pupils. It includes an invitation to them to suggest a particular target they need to focus upon and an indication of how their work will be assessed.

HUMANITIES DEPARTMENT
FARMING

WHAT YOU WILL DO IN THIS UNIT

1. Visit a local farm
2. Contrast this farm with a hill farm using photographs, books and a video.
3. Use new geographical words to describe different types of farming, their locations and how much space they occupy.
4. Use and draw diagrams concerning farming year, inputs and outputs, crop rotation etc.
5. Use and draw plans of farms.
6. Consider the changes in farming.
7. Discuss the environmental issues concerning farming.
8. Link the distribution of different types of farming in the British Isles with climate etc.

ASSESSMENT

1. Concept map.
2. A chart to show the contrast between hill farming in The Lake District with arable farming in East Anglia.
3. Poster to show conflict over use of land for farming.

SKILLS TARGET

Listening carefully and asking good questions

YOUR TARGET

Figure 7 Sharing planning with pupils (developed by Vicky Farthing, Beyton Middle School)

Short-term Planning

Detailed plans for individual lessons or activities are unlikely to be required where geography is taught by specialists and well-developed schemes of work are in place. However, for teachers whose experience or confidence in delivering

ACTIVITY REFERENCE:	Geography	CLASS:	GROUP	TERM AND YEAR:

ACTIVITY PLAN: Identifying similarities and differences between buildings and their uses.

TEACHING STRATEGIES	CURRICULUM REFERENCES	MORE IDEAS
1. Pupils watch video of 'Journey down my road', which shows different types of buildings in immediate vicinity - schools, houses, shops, etc. Pupils identify buildings, consider use and similarities/differences to other buildings on video.	**Geography** - identify features of and talk about places in their local area - investigate the uses of buildings	1. Pupils to take photos or make videos of their street or street near school.
2. Give one photo to each pupil and ask them to identify which type of building it is and sort the photos into trays according to type eg. houses, schools, churches, hospitals, garages and train stations.	- communicate about work and leisure activities - identify and name familiar features, eg. buildings, parks - use pictures and photos to identify features, eg. homes, railways, etc	2. Grid reference the boards and place the buildings according to the grid reference.
3. Give out symbol cards of same types of building. For each card the pupil identifies type of building and then matches to photo of same type of building on pinboard.	- identify similarities and differences in the ways land and buildings are used	3. Use of grid reference to identify their own street/road.
4. Give out boards with street running across and divided into four sections. Small group of pupils each with board take turns to throw 'buildings' dice (with symbols of buildings on it), and select a photo to match the building thrown on the dice.	- use geographical terms, eg. road, house, shop, etc - make plans of actual or imaginary places.	4. Video in rural/urban area and discuss differences.
5. Pupils decide on which section of board to place photo.	**English** - discussion with others - listening and giving weight to others' opinions	5. Use work on quadrants placing photos on street to start simple classroom maps using a sheet divided into four and a floor area divided into four using tape.
6. When each board is complete, get them to describe their streets and discuss use of buildings and anything relevant to their layout, eg. if the house is here and the school on the other side of the road here, what will the children have to do to get to school.	- talking and listening in the group - articulating personal feelings expressing opinions - turn-taking - use of symbols	
7. Draw a plan of where they had placed the buildings.	**Science** - encourage sorting and grouping of objects, etc. in their environment, using their senses; noting similarities and differences.	
8. Ask them what else they would like to have in their street and why?		
Activity differentiated through use of simplified boards and real objects.		

COMPLETED BY:	DATE:

Figure 8 An activity plan on the use of buildings

the subject is limited, more detailed planning may assist in providing guidance and in reviewing and evaluating teaching effectiveness and pupils' progress. For geography co-ordinators in primary and special schools, a resource bank of individual activity plans will be useful to other teachers in the school and will provide a focus for staff discussion and development. These resource banks may categorise the activities according to the five areas of geography (skills, places, physical, human, environmental) described in chapter 1, although in practice most activities will integrate several areas.

There are many different formats which could be used for lesson plans. The example given in figure 8 is of a lesson (or series of sessions) on the use of buildings for pupils with severe learning difficulties. It provides a description of the procedure followed in the lesson, references to the National Curriculum programmes of study and some ideas for related activities arising from the lesson. These activities are referred to again in chapter 3.

Rose (1994) describes the development of curriculum modules which provide lesson outlines, curricular referencing and a list of the resources required. Each module is like a scheme of work, consisting of a series of lessons. An example of part of a module on 'Learning about maps' is provided in figure 9.

DIRECTION AND MAZES

This lesson looks at directions forwards, backwards, left and right. It also aims to encourage pupils to think about direction of travel.

Equipment

A selection of PE apparatus, a long rope, computer, software, worksheets (on mazes), paper, pencils, felt tipped pens. Directions board game with dice and counters.

Activity 1

a) Make a rope trail around the PE apparatus in the hall so that pupils have to follow this in, out, under and over apparatus. When they have been around successfully, let them attempt to negotiate the same course in pairs. Pupils can be joined together using PE ribbons for this activity. Communicate with pupils about the directions in which they are travelling.

b) Using PE apparatus, create a maze in the school hall. Direct pupils through the maze using instructions such as walk straight on, turn right/left. Get pupils to direct each other through the course. If there are pupils who are confident enough, try blindfolding them and directing them through. **Please be aware of safety factors if you try this.** Allow plenty of space for pupils in wheelchairs. Some pupils may be encouraged to push pupils in wheelchairs around the course.

Activity 2

Use the directions board game with pupils. Communicate with the players about directions. This game encourages pupils to follow directions and instructions.

Activity 3

Use the simple maze worksheets. Encourage pupils to draw their route through the maze and to show this on the worksheet. Pupils may like to try designing their own maze, either on paper, or using PAINTSPA (on the Nimbus computer) or GRIDIT (on the Archimedes computer - see chapter 4 of this book for an example).

> **Activity 4**
>
> Use ANIMATED ALPHABET on the computer which has a simple maze.

> **Curriculum References - Geography**
>
> Pupils should be taught to follow directions, including the terms up, down, on, under, behind, in front of, near, far, left, right, north, south, east, west.
>
> Pupils should be given opportunities to apply and develop their IT capability in the study of geography.

Cross references to other curricular areas are given at the back of this module.

Figure 9 Module format for a lesson from Wren Spinney School, Northamptonshire.

Progression in Geography

The National Curriculum Council (1993a & b) suggest ways in which planning can ensure progression in geography. It involves sequencing learning to enable previous experience to be built upon and repetition to be avoided. Progression can be planned for through increasing the level of difficulty of the task, the breadth, the depth, the complexity of the concepts, the range of scales, moving from concrete to more abstract concepts and increasing the range and complexity of geographical vocabulary. These methods of developing progression will be used in different combinations across the key stage.

Recording

The fundamental decision to be taken in relation to recording in any subject is to identify the purposes for which it is required. Currently, teachers are required to report progress on all National Curriculum subjects and assessment and recording should provide the evidence for that progress. In addition to meeting the reporting requirements, recording of progress should be used to inform subsequent planning and teaching. However, recording of achievement in geography appears to be poorly developed (OFSTED, 1993b).

The overload in the 1991 curricular requirements led to wide scale adoption of checklists based on the statements of attainment. The judgements made were of variable accuracy. Traditionally, some schools developed extensive checklists for recording pupil progress or bought the commercially available schemes which were developed in response to the National Curriculum. While these schemes may be helpful to clarify issues concerning progression, they are of limited use on a daily basis as they are too time-consuming and cumbersome and provide more information than is likely to be used. Some staff working with pupils with learning difficulties have been so concerned with the slow rate of progress that any response given by the pupil is noted and sometimes a cross on a sheet is given to record the absence of a response. This creates extra demands on staff and results in more information being generated than is necessary. The revised curricular requirements, while demanding more from teachers in terms of development work on assessment and recording, will provide a more realistic basis on which to make judgements.

Another purpose of recording can be to provide sufficient information to indicate coverage of the curriculum, which will highlight what learning opportunities or experiences pupils are offered. This will provide the information needed for an annual curricular audit. In theory, the schemes of work could provide the basis for keeping a running record of coverage but, in practice, intentions and plans do not always materialise so it may be necessary to note on the schemes of work which areas were not covered. A register of attendance would then be sufficient to indicate which areas were not covered by an individual pupil due to absence.

Recording progress in geography can be tackled in a number of ways. The school should have a whole school policy for assessment, recording and reporting which will provide the basis for recording in geography. It may be helpful to use the areas of geography described in chapter 1 as the basic categories against which pupils' progress is noted.

It is important to distinguish clearly between evaluation of the activity in terms of teaching and progress made by pupils. In the Framework for Inspection (OFSTED, 1994), separate criteria are given for quality of teaching, quality of learning and standards of achievement. Teachers must note progress (quality of learning) and evidence of standards indicated by responses or pieces of work.

A system which enables pupils' responses to be recorded **if and when they are significant** is needed. For some activities, one sheet may be needed for each activity, for others, salient responses from several sessions can be recorded on the same sheet with the date of each session noted. Decisions regarding what constitutes a salient response should be made in the light of individual pupils' priorities for learning and progress in the pertinent areas of work in geography. For one pupil, this might mean recording every time she responds to directional instructions, whereas for another pupil a significant response might occur the first time she identifies a difference between the locality of the school and a contrasting locality.

Assessment

An on-going recording system may provide important information for the assessment of pupils' progress. However, it is possible that some pupils will have understood more than they are able to demonstrate within the activities provided. Hence, a system of assessment which enables the teacher to check that the pupils' progress has been accurately monitored is essential. This might involve targeted observation of individual pupils, questioning or listening to pupils and evaluation of examples of work or responses noted in the records. It is important to build in opportunities for assessment at the planning stage.

The management of assessment is central to its effectiveness. OFSTED (1993a) noted that inconsistencies in practice by teachers in the same department pointed to the absence of agreed policies for assessment. Pupils' progress on other areas of the curriculum should also be noted although responsibility for the assessment relating to one subject within a lesson addressing another subject remains a much debated area of concern at secondary level when pupils are subject rather than class based.

In assessing achievements in geography with pupils who have learning difficulties, it is important not to assume that their profile will be even across different skills in geography. For example, a pupil able to link goods to the shops which provide them may not be able to follow a simple map, even when the two tasks are presented at comparable levels. Furthermore, some pupils, for example, pupils with emotional and behavioural difficulties, may demonstrate different levels of achievement depending on whether they are assessed during a group or individual activity. The context in which tasks are assessed will be important.

The publication *Children's Work Assessed* (School Examinations and Assessment Council, 1993) provides examples of pupils' work which illustrate how assessment could be used to inform teaching. Effective progression depends on accurate assessment. It is critical that appropriate types of information are recorded to provide a basis for assessing progress. Marking of work, for example, does not always provide diagnostic information clarifying for the pupil and teacher the improvements that are needed. Furthermore,

WREN SPINNEY SCHOOL

This is to certify that

Completed the module

LEARNING ABOUT

MAPS

During the Summer Term 1993

Head Teacher   Chair of Governors

Side 1: Record of Experience

LEARNING ABOUT MAPS

I Can...
Find places on a map

Identify the countries of the British Isles on a map

Recognise five map symbols

Use a simple grid reference

Follow simple directions

Cooperate with friends to make a map

Follow a trail around school

Find five rooms on a map of the school

> **A: Without help. B: With some help
> C: With a lot of help**

This LEARNING ABOUT MAPS MODULE has been moderated by:

Inspector
Northants Inspectorate and
Advisory Service

Side 2: Record of Achievement

Figure 10   Distinguishing experience from achievement: an example of accreditation (Rose, 1994, p.62)

recording often notes curricular coverage rather than the achievement of individual pupils, making it impossible to use as a basis for assessment and thereby for planning future work. The accreditation system used by Wren Spinney School in Northamptonshire (Rose, 1994) reproduced in figure 10 provides an excellent example of how curricular coverage and specific individual achievement can be recognised separately.

The Role of the Co-ordinator

Overall responsibility for planning, recording and assessment will fall to the subject co-ordinator. Their role is sufficiently demanding and important to suggest that identifying a member of staff with genuine commitment to the subject is a priority. Experience of some primary and special schools is that geography is not seen as a priority. Underdevelopment of the subject suggests a need for a stronger co-ordinator than might be needed for a better established subject, since the teacher will be required to 'sell' the subject to other members of staff and demonstrate through example its potential for learning. The identification of a co-ordinator should not be seen as a justification for all the other teachers in the school to ignore the subject altogether.

Possible responsibilities for the co-ordinator include:

- drafting of the policy statement, long-term plans (across key stages) and schemes of work;
- planning for progression and differentiation;
- auditing current curricular coverage;
- developing a good understanding of the curricular requirements;
- creating resource 'boxes';
- developing a range of appropriate activities for others to use;
- monitoring progress across the school;
- listing and acquiring resources for teachers;
- taking responsibility for the work of other teachers who teach geography;
- ensuring geography is prioritised on the school development plan when appropriate;
- providing staff development through team teaching, staff meetings, etc.

In this chapter a range of methods for planning, recording and assessing in geography has been presented. There is no one model which will suit all schools and all pupils. However, there are some basic principles which are important to use as a guide to developing effective practice. These are:

(1) Ensure the planning system in the school provides a broad and balanced curriculum which is planned sufficiently far ahead to ensure continuity and progression.

(2) Include opportunities for assessment at the planning stage rather than bolting them on as an afterthought.

(3) Ensure that the primary reason for assessing pupils is to identify what they have learnt so that the teacher knows what to do next.

(4) Adopt a recording system that enables pupils' progress in geography to be noted in sufficient detail to inform assessment but does not generate

volumes of unused information.

(5) Assess pupils through a variety of methods to ensure they can demonstrate what they have learned.

(6) Involve pupils where possible in planning, recording and assessing.

(7) Define a clear, realistic role for the co-ordinator and ensure adequate support is provided.

These principles should assist in developing a coherent structure for planning, recording and assessing in geography. The next chapter provides examples of geography teaching from schools and local authorities, undertaken with pupils whose needs are very diverse.

CHAPTER 3

Practical Activities and Resources for Teaching Geography

Introduction

The purpose of this chapter is to demystify geography teaching further for any readers who still feel perplexed and to stimulate ideas for exciting and effective teaching to meet the diverse needs of *all* pupils in schools. The activities and resources presented in this chapter are largely the work of other teachers, local authority advisory staff or have appeared in other publications and are acknowledged as such. They are not presented as 'perfect practice' judged against a set of predetermined criteria. Each has strengths and weaknesses but all can stimulate further developments in geography teaching. If they promote learning in pupils who have been regarded previously as failures, difficult to teach or not able to benefit from geography, they will have served their purpose.

The range of possibilities and opportunities for teaching geography may have increased through the introduction of a National Curriculum which has stimulated a broader understanding of the subject. Teachers of pupils who demonstrate limited literacy skills will need to plan creatively and flexibly to ensure the greatest possible access to geography. Methods of increasing access to geography are discussed more fully in chapter 4 and in Sebba (1991) and Sebba and Clarke (1993).

This chapter describes geography activities from key stages 1, 2 and 3 which have been used by teachers working with mixed ability groups and groups of pupils with learning difficulties. They are presented under the headings introduced in chapter 1: geographical skills; places; physical geography; human geography; environmental geography.

The decision to use these categories is somewhat arbitrary. Most activities cover more than one skill or area of knowledge. However, these areas represent the essential learning involved in geography and are likely to remain a focus for teaching whatever revisions or changes are made to the curricular requirements in the future. It is hoped that the use of these categories will encourage teachers to consider which aspects of geography they are targeting each time they develop an activity and to look at the overall breadth and balance of activities planned.

The activities address a range of content suitable for pupils of different ages. Where possible, content involving more difficult concepts is presented in ways which might simplify the ideas. However, rigorous assessment will be needed to ensure the teacher is fully aware of which concepts individuals have understood. The same content often provides opportunities for some pupils to learn one concept while others learn another. This is further explored in the section on group work in chapter 4.

Geographical Skills

Geographical skills include map work, field work and the enquiry skills involved in these and in the subject more generally. Many of these skills, for example observation, are not exclusive to geography. The examples here include work on plans, maps, grids and trails. They are designed to develop an understanding of direction, representation, scale, orientation, interpretation of symbols and measurement. While opinions vary as to the order in which map work skills are acquired, it is clear that very young children can develop some basic skills. Catling (1984) provides one possible model for progression of skills in map work. A clear and very readable account of map work, and of the other areas of learning described in this chapter, is provided by Wiegand (1993). The *Mapstart* series (Catling, 1993, revision), *Start Orienteering* series (McNeill and Renfrew, 1990), *Time and Place* (Harrison and Harrison, 1991), *Discover Maps with Ordnance Survey* (Harrison and Harrison, 1988) and Suffolk County Council (1991a) guidance booklet provide excellent classroom resources on map work and orienteering. The Geographical Association (address in Resources list) also publish many relevant books for teachers. The journal *Special Children* produces resource packs with each issue, many of which address geographical content. One published in January 1991 was about 'Finding Your Way' and included ideas on mapping skills.

The use of compasses, globes and atlases also contributes to these areas of learning though not specifically illustrated here. Some mention of these is made within activities described within other sections of this and the next chapter. Field work and enquiry skills are drawn out within each activity, as appropriate throughout chapters 3 and 4. It is important to note that for pupils with visual difficulties, the Royal National Institute for the Blind (RNIB, address in Resources list) distribute tactile compasses, atlases and globes to enable these pupils some access to their use. Tactile maps are also available from various sources. However, for pupils with substantial learning difficulties and additional visual disabilities, activities addressing geographical skills remain difficult to access.

Map work

Map work can begin with activities designed to develop an understanding of a plan, at the simplest level, by identifying everyday objects from their outline, and then from the outline created by drawing around them, or using an overhead projector to obtain images. Inspired by *Mapstart* and *Time and Place,* Carol Doran (1992) developed an extensive set of classroom resources which she used with eight year old pupils with severe learning difficulties and reported responses on a wide range of geographical skills. She started by inviting the pupils to make textural rubbings of features in the playground, including drains and paving. The main skill being developed was the idea that to produce a plan you have to look down on to the object.

In the second session each pupil had a sheet of paper divided into four with one object in each quadrant. They drew around the objects and were then invited to answer questions using their plan, such as 'Where is the square shaped object?'. Out of their vision one object was removed and using their plans they were asked to identify which object was missing. In the next session directional terms were introduced – left, right, top and bottom. Again an object was removed and the pupils had to describe, using two of the terms, from which quadrant the object had gone. Some of the pupils were unfamiliar with the terms left and right and a sweet was then hidden under one of the objects to increase motivation.

Photographs of familiar objects in the classroom taken from a 'side view' and a 'plan view' were given to pupils who were asked to identify and find the object. Each pupil then matched the position of the object to the view they had. By the end of this session they were all able to pick out the 'plan' or 'picture' when asked to do so. The use of photographs of familiar objects generated great interest so the teacher decided to make more use of these. She prepared a large 'plan' of the classroom with photographs of different features of the room. Each pupil was given a photograph which they had to identify and use to lead the rest of the group to it. This latter activity was less successful, with other pupils taking the opportunity to wander off elsewhere! However, the classroom plan was used to teach the pupils about the use of a key, identification of symbols and finding the relevant feature on the plan. The teacher was surprised to find that all the pupils were able to do this, one demonstrating an ability to use the key before the teaching had begun. Some pupils were able to draw routes on to the plan, which had been laminated to enable this usage.

Trails, route-finding and mazes

The 'String Trail' which I described elsewhere (Sebba, 1991) also involves the use of maps. A piece of string is laid down around the outside of the school building. At a number of points on the string different coloured pieces of tissue paper are attached to the string. Small groups of pupils (3-4 maximum) are presented with a 'map' of the school as shown in figure 11 and a set of coloured pens or crayons to match the colours of tissue used. They are asked to follow the string around the school until they get to a piece of tissue paper. They try to locate where they are on the map and mark this using the matching coloured pen. At the end of the trail they join up the coloured marks on the map to show where they have been and discuss their route.

It is important in this activity that the groups of pupils start at different points on the string or travel in different directions so as not to distract each other. As shown in the figure, the maps can be traditional architect-type plans (a), or maybe simplified using symbols or photographs (b) according to the different needs of the pupils involved. This activity has been done in many primary and special schools and pupils appear to enjoy it as well as responding well. For pupils who have profound and multiple learning difficulties the activity may not elicit any geographical skills but may provide the opportunity

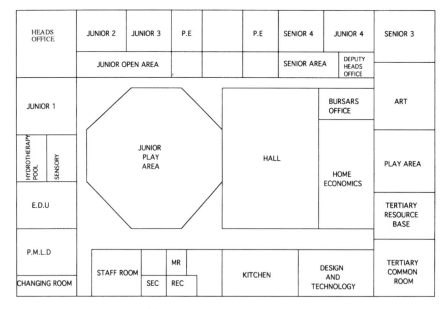

(a) WEST OAKS SCHOOL GROUND PLAN

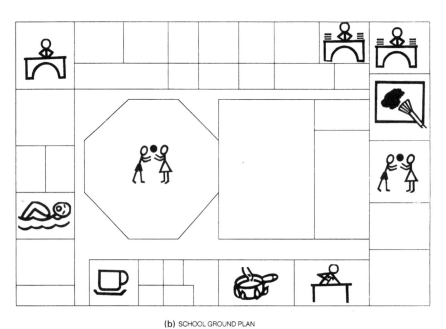

(b) SCHOOL GROUND PLAN

Figure 11 Examples of maps used in the String Trail, from West Oaks School, Leeds

to pursue other priorities such as mobility and communication. Following the string, matching the colours or co-operating with others may be responses noted rather than the map work skills involved, in which case geography has provided the context in which learning has occurred in another area of the curriculum. The recording will need to distinguish this precisely.

It is possible to make this activity much more demanding by asking pupils to identify errors in the map, giving them a map that requires completion, or asking them to construct their own map for others to follow. It is also an activity that lends itself very well to the enhancement of group work skills, in particular 'jigsawing', which is fully discussed elsewhere (Sebba *et al*, 1993).

At Windsor School in Clacton, the pupils were invited to use a map of a street in the town to find their way and answer some questions on a symbol-supported worksheet as shown in figure 12. As in the string trail, this map could be constructed using photographs or computer-generated symbols as described in chapter 4, in order to increase access to the activity for specific pupils. It is relevant to the study of the local area within places and to many aspects of human geography such as why journeys are made, why shops are located in particular places, and the provision of goods and services.

Mazes provide a useful context for teaching route-finding and directional skills and provide strong links with mathematics. Within a PE lesson an obstacle course can be constructed in which pupils follow mats, tunnels, steps and benches to reach a particular destiny. David Banes at Meldreth Manor School used rolls of wallpaper to create paths in the hall for pupils who crawled, rolled, bottom-shuffled or used wheelchairs to follow the paths. Soft play equipment can be particularly useful as it enables paths to be created in which the surrounding area becomes obscured. These activities can be adapted for each pupil, for example, using tactile paths for the pupil with a visual disability. Further activities with computer-generated mazes are described in chapter 4.

Grids and co-ordinates

The classroom resource material listed at the start of this section on map work include many suggestions of ways in which co-ordinates can be introduced. The commercially available game 'What's in a Square' provides a basic introduction to the concept. The Humberside County Council (1992) material on geography and learning difficulties includes an activity called 'Square Routes' illustrated in Figure 13. Each pupil has a sheet with the faces on it, preferably made from their own self-portraits. The worksheet requires the pupil to follow directions so that one 'face' meets another one. They can simply move their fingers from one square to the next or the faces could be movable on a fixed board.

I have adapted this activity to enable pupils for whom worksheets are not appropriate, to participate in it, sometimes within a PE lesson. Each pupil stands or sits in their wheelchair in a carpet square or on a mat or inside a hoop. In turn, they are asked to move 2N, 1E, for example, and report who they meet. The directional terms used can be varied for different pupils, for example using terms such as towards the window, towards the door or forwards, backwards or by designating the walls different colours with paper or balloons. Pupils can take it in turns to give the instructions. Within the PE session, games such as 'Sharks' can be used to reinforce directional skills.

Harrison and Harrison (1991) describe some useful grid reference games suitable for young children or older ones with limited literacy skills. One involves reconstructing a picture while another one uses two spinners, one with

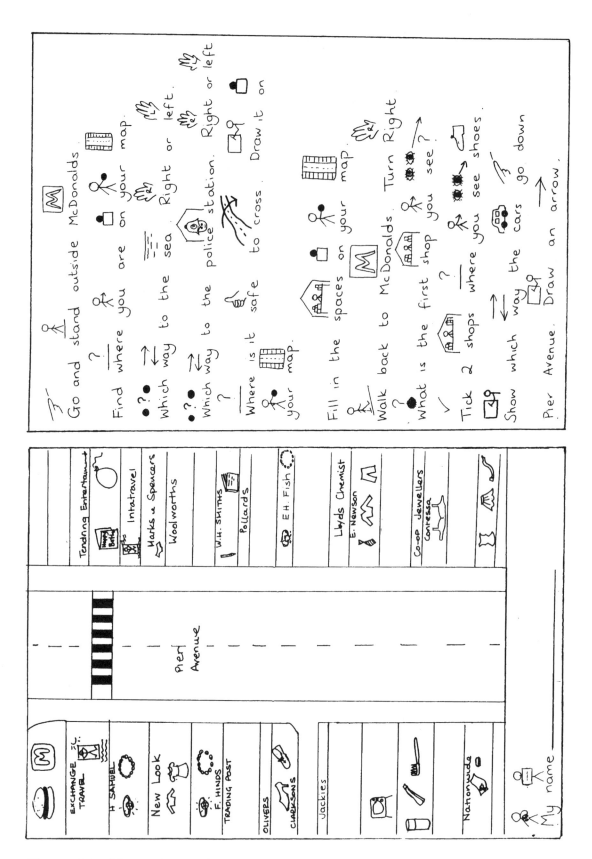

Figure 12 Map and part of a worksheet from Windsor School, Clacton, Essex

34

Figure 13 Square Routes from *Access to Geography* (Humberside County Council, 1992, p. 115, reprinted with permission)

the labels from one axis (a, b, c, etc.) and one with the labels from the other axis (1, 2, 3, etc.). The spinners then determine which square the child has to find and mark off, the object of the game being to complete all the squares. We have devised our own grid games in schools which tie in with the work they have been doing in geography. Thus, in one school a project on the woods and another on use of buildings included grid games designed to enhance the concepts involved as illustrated in figure 14.

Geographical skills through stories

Potentially, all areas of geography can be enhanced through the use of stories. These may be read by the teacher or pupil, retold by the teacher or pupil, illustrated, incorporated into work in other curricular areas or used as a basis for introducing a new concept. There are an increasing number of reading resources being developed for older readers with more basic reading skills. Many of these schemes include books which are pertinent to areas of geography such as seasons, transport, houses and homes, places and environmental issues.

Many of the resources for teachers include suggestions for use of stories in geography. Some local authority advisory services have produced this type of guidance (for example, Suffolk County Council, 1991b), and two booklets (Routh and Rowe, 1993; Rowe and Routh, 1992) aimed at key stages 1 and 2 list and annotate many appropriate stories under the same areas of learning as in this chapter. These publications all suggest stories designed to enhance geographical vocabulary and mapping skills. Well-known examples such as *Winnie the Pooh, The Jolly Postman, Thomas the Tank Engine, Little Red Riding Hood* and *The Three Little Pigs* all involve journeys which can be mapped out.

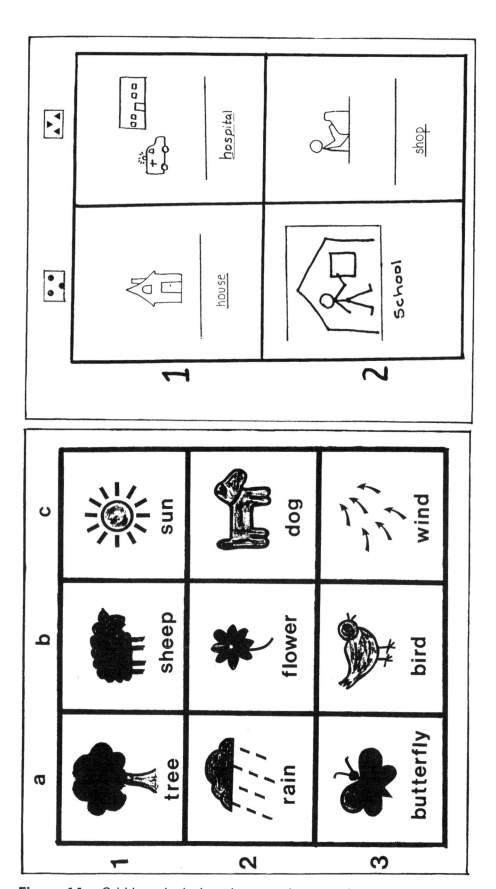

Figure 14 Grid boards designed to complement other geography work

However, while stereotyping is clearly evident, the 'Postman Pat' series appears to include more geography than the average geography text book! Here is an extract from *Postman Pat's Tractor Express* (Cunliffe, 1983) in which I have highlighted all the geography coverage to illustrate this point.

> Every **summer, visitors came to Greendale**, to walk in the **hills** and camp in the **meadows**. "What lovely **weather** for people on **holiday**," said Pat, as he drove along the **valley**. He called at the **village post-office, to collect the day's post.**

Thus, in this short extract, references are made to seasons, weather, work/leisure, where people go on holiday and why, services (post) and physical features (hill, valley, meadow). It can be useful to introduce a new area of geography with a familiar story, enabling a concept to be grasped within a familiar context before considering it more widely.

Places

An excellent exploration of teaching place knowledge and about localities is provided by Wiegand (1992; 1993). In these publications he provides a convincing review of the evidence supporting the need for the study of places, and, in particular, of distant places. He suggests that education for international understanding is critical in pupils' development. The study of the immediate environment needs to be balanced with learning about national and international contexts. He also argues that since children are more interested in people than places, the traditional approach to teaching geography, which looks at the physical characteristics of a place and then at the effects these have on lifestyle, should be reversed, so that lifestyle is followed by an attempt to explain the factors which contribute to it.

The information pupils have about distant places is frequently biased and stereotyped, for example suggesting that people in developing countries all live in mud huts. These distorted images are all too frequently promoted in geography textbooks and resources. At best, pupils' information is highly selective based on a particular aspect of distant places rather than a rounded understanding. Hence, Wiegand (1992) quotes the example of distant places known about by children aged 7-11 years, in which 'Zimbabwe' featured for two main reasons: the children liked saying it and Liverpool F.C.'s one-time goalkeeper was born there! Stereotyping is a particularly prevalent issue with pupils with learning difficulties, who, without wishing to stereotype them, sometimes have more limited experiences, particularly if they have substantial difficulties.

Wiegand (1992) suggests ways in which teachers can tackle stereotyping by providing more detailed information about groups of people which fragments the global stereotype. This flooding of information might be less effective with pupils with whom the amount of input needs to be limited. Another approach he suggests, which may be more effective with all pupils, is to deliberately challenge one or more attributes on which the stereotype is based. This is a technique my colleagues and I frequently use with teachers who occasionally express stereotypical attitudes in relation to pupils labelled as 'autistic', 'dyslexic' or 'Down's syndrome'. By presenting evidence which directly challenges an attribute associated with the label, confidence in the stereotype is reduced. In the context of geography teaching, it is important to present a broad range of information about life in distant places and to explore the attitudes and feelings of a wide variety of people in those places.

Study of places should involve as broad a range of resources as the teacher can find or develop. These might include pupils' own experiences from travel or visits, videos, television and films, books, photographs, postcards and drawings, maps, objects including food and links with other schools. Field work has an important role here in the study of the locality and possibly in the study of a contrasting locality in the UK. Some schools have successfully tackled a contrasting locality beyond the UK by arranging a school trip or holiday which has supported work in geography and modern foreign languages simultaneously.

Local studies

Carol Doran (1992), some of whose work was described in the previous section, developed a series of photographs relating to the school and the route from the school to the park. The pupils used the photographs to retrace their journey and to reinforce the identification of features they had observed, for example, gate, pond and bridge. Word cards were also used with those pupils for whom these were appropriate. Examples from her resources are illustrated in figure 15.

The *Our Town* activity book (Hawkin, 1982) provides an example of how basic information about an area can be collected and a local study conducted. Each pupil or group of pupils can construct a booklet about their own locality including drawing maps, looking at the use of shops in the area (see figure 22 in the section on human geography), traffic surveys, road signs and distances. It might be possible for pupils to provide one another with contrasting features where pupils' homes are widely dispersed. Sensitive handling of the concept of 'home' may be needed where pupils have distressing home lives or live in particular circumstances, such as travellers.

Local studies with a specific focus

The use of buildings provided the focus for a local study in one special school in which my colleague, Sandra Galloway, and I were working. Sandra made a video of the street outside the school going a sufficient distance to take in a wide variety of buildings – detached, semi-detached, bungalows, high rise flats, garages, churches, shops, schools, a hospital and a station. She took photographs of the same buildings and prepared a set of 'symbol' cards to match. The pupils undertook a series of activities based on this material, the plan for which appears in figure 8 in chapter 2. They observed some of the buildings directly, although the field work was limited by the distance that would need to be covered in order to observe the range of buildings required and the sensitivities involved in a whole class standing outside someone's house apparently invading their privacy.

The pupils watched the video and, while the picture was paused, identified the type of building and discussed similarities and differences between buildings. The photographs and symbol cards were sorted and matched

We walked over the...

bridge

We passed a ...

house

Figure 15 Photographs and wordcards used in a local study (Doran, 1992)

according to various criteria. The number of cards and criteria varied to suit different pupils' needs. Some pupils played a game in which they each had a board that represented a street and they threw the dice with symbols of buildings on it. They selected the photograph, symbol card or a three dimensional model of the type of building shown on the dice and decided where to place it on their card. When the pupils' cards are complete, they are

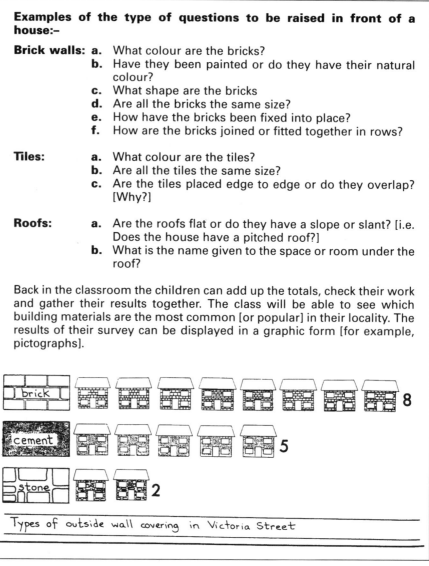

Examples of the type of questions to be raised in front of a house:-

Brick walls:
a. What colour are the bricks?
b. Have they been painted or do they have their natural colour?
c. What shape are the bricks
d. Are all the bricks the same size?
e. How have the bricks been fixed into place?
f. How are the bricks joined or fitted together in rows?

Tiles:
a. What colour are the tiles?
b. Are all the tiles the same size?
c. Are the tiles placed edge to edge or do they overlap? [Why?]

Roofs:
a. Are the roofs flat or do they have a slope or slant? [i.e. Does the house have a pitched roof?]
b. What is the name given to the space or room under the roof?

Back in the classroom the children can add up the totals, check their work and gather their results together. The class will be able to see which building materials are the most common [or popular] in their locality. The results of their survey can be displayed in a graphic form [for example, pictographs].

Types of outside wall covering in Victoria Street

Figure 16   Example of questions and display of results from a local housing survey from *Ourselves* (Simkin, 1990, p.18)

invited to describe their street and discuss the use of buildings and anything relevant to their layout. For example, one pupil noted that if the school was on one side of the road and the house on the other, the children would have to cross the road to get to school. They also considered what else they would have liked in their street and why.

We attempted to meet the needs of a pupil with visual impairment and learning difficulties in this group through the use of the models, but conceptually these did not represent her experience of buildings and were of little use. A more effective approach might have been to have produced tactile symbol cards and tactile dice. Similar resources to those developed in this activity could more effectively use an Ion camera enabling stills to be projected through the screen and reproduced in a variety of ways. Extensions of this activity could include developing similar resources relating to a contrasting area, for example rural/urban, perhaps one of the streets on which a pupil lives in a school with a wide catchment area. The software package 'Urban studies' (Science Education Software, address in Resources list), reviewed in the next chapter, enables pupils to construct a graphical profile, to scale, of the buildings in a street and print it out for display (illustrated in figure 31 in the next chapter). This could be used within a local study on buildings.

The Tressell pack *Ourselves* (Simkin, 1990) includes a description of a series of activities involved in a local housing survey. One of the activities involves observing and recording what materials are used to make the roof and outside walls of houses in the immediate vicinity. Possible questions that could be asked during the field work and an example of how the information could be displayed are given in figure 16.

The Tressell resources go on to suggest activities on building materials and an investigation of how bricks are made which is relevant to environmental geography, history and technology.

Distant places

Resources to be used for studying distant places will need to be carefully chosen in the light of the discussion above about stereotyping and to ensure accessibility in terms of literacy demands. Classroom texts which address more advanced geographical content with a simplified text are particularly helpful for this work. Wayland publishers have an information series which includes books on conservation, energy, volcanoes, weather, festivals and Europe. They have been rewritten using shorter, simplified text with the same illustrations and concepts. The example in figure 17 is from *Life in Europe* (Elliott, 1994).

Teaching pupils about distant places can be easier if someone in the group or one of their close relatives has been there. In figure 1 in chapter 1, the example of the enquiry process used holiday postcards as a starting point. A similar activity is described in the Wren Spinney School (undated) Module on 'packaging and advertising'. Here pupils are encouraged to examine examples of holiday brochures and brochures about their home region, Northamptonshire. Pupils discuss holidays they have been on and previous school residential trips. They consider what they like to do on their holidays and look to see if they can find these activities in the brochures.

Pupils are encouraged to select a holiday destination and compare features of their chosen place with those of Northamptonshire which are displayed on a chart. Finally, they make a poster advertising Northamptonshire as a holiday destination. In this module, pupils are being given opportunities to learn about features of the environment, preferences, weather, leisure, transport and similarities and differences. The activity focuses on their own locality (making learning more meaningful) but provides contrast with another locality. Enquiry skills are promoted throughout the activity.

Towns in Europe are growing day by day as more people come in search of work, and as businesses and industries develop.

House prices are very high which means that fewer people can buy their own homes. However, rents for flats, houses and offices are also high.

The governments of many European countries are looking at ways to make life better in towns and cities.

▼ Look at the chart below. Can you see that more Germans than Britons live in flats? More Britons own their own homes.

Who lives in flats

64 per cent

21 per cent

Great Britain   Germany

People who own their own homes

58 per cent

37 per cent

Great Britain   Germany

Figure 17 A sample page on towns and cities from *Life in Europe* (Elliott, 1994, p. 29, reprinted with the permission of Wayland (Publishers) Ltd)

The *Time and Place* materials (Harrison and Harrison, 1991) include a similar activity in which pupils' postcards and holiday photographs form the basis of an activity focusing on the world. Pupils are invited to display their pictures on the world map. They are also invited to display family links through photographs of relatives living in different countries, clothes and objects from different parts of the world, and food which indicates on the packaging from where it originates. Further discussion could focus on transport or climate.

Linking schools

An excellent method of developing resources on a contrasting locality, in or beyond the UK, is to form a link with another school, preferably in a contrasting area, although the contrasts could be more subtle for pupils with less substantial learning difficulties. George Hastwell School on Walney Island in Cumbria has established an electronic mail link with a small school on the island of Rovaer in Norway. The schools exchanged maps, postcards, photographs, brochures and answered questions generated by the pupils as shown in figure 18. This necessitated the pupils collecting substantial information about their own locality and deciding what it was important for people to know about the area. In addition, pupil profiles were exchanged and

```
Item     1596293              25-Jan-93      00:53PST

From:    NOR0091              Rovar Skole,ROVAR,NO,AGE

To:      HASTWELL.G           George Hastwell School,GB,IMU

cc:      HART.B               Imagination Technology,GB,IMU
         AGE.RD$              Apple Global Education R&D

- - - - - - - - - - - - - - - - - - - - - - - - - - - - - - - - - - - - - - - - - - - - -

Sub:     Re: QuestionsAboutRoevaer

*************************************************************************
**********
NOR0091 RØVÆR SKOLE, NORWAY
*************************************************************************
**********

Dear Class 4 at George Hastwell School,

Her come our answers to you questions! Sorry for the delay.


*How big is your island?
   The island is 1,4 square kilometers. (!!!)

** Do you have lots of boats?
We have 2 big fishingboats on the island, about 15 other smaller
fishingboats and our ferry.


*** How far away are you from the main land?
   About 10 kilometers.

**** How long does it take you to get to the main land?
   It takes 45 minutes by our ferry.

***** Does your island have some shops on it?
   Yes, the island has a small shop where you can buy the things you
need.

****** Does your island have a swimming pool?
   No, we swim in the sea (in the summer, of course!).

******* Are your houses made out of wood or are they made out of
bricks? Most houses are made  out of wood; but none are made out of
bricks.


******** Do you have a golf course on your island? No, we don't have
any golf course. There is none in Haugesund, either.

********* Our weather has been very stormy lately.  What is your
weather like at the moment?

At the moment we have a north-western storm with some rain and sleet.



                        AppleLink:HASTWELL.G      Page 1
```

Figure 18 Answers to pupils' questions about a contrasting locality using an electronic mail link, from Clare Martin, George Hastwell School, Cumbria

the school in Norway provided information for a project that George Hastwell pupils were doing in history on the Vikings. The pupils prepared their own report on the project, supported by the teacher, which is communicated through symbols and words, making it more accessible than the information in figure 18. A section of this report appears in figure 32 in the next chapter.

The school has exchanged electronic mail messages with schools in Portugal, Canada and USA and the location of the schools is marked on a world map on the classroom wall. The class researched how long it would take them to travel to each place.

Economically developing countries

Where available, materials and resources from the pupils' families about developing countries can be used. However, many teachers will be dependent on published resources for work on economically developing countries. The development agencies publish some useful materials, the most widely used being 'Chembakolli', which is a resource pack about life in a village in India published by Action Aid (address in Resources list). The materials include a map, statistics, maps and graphs about India, photographs, pupil's activities and teacher's notes. A separate set of slides is also available. One set of picture cards in the pack traces the life of Chanda, a woman from the village. A successful activity observed in St Joseph's primary school in Sudbury, Suffolk, involved pupils making a book of a day in the life of their own mother as a comparison. Other examples of the use of this resource pack are described by Brace (1992).

Other sources of appropriate material include video, travel guides, brochures and books. Stories about life or events such as floods or earthquakes, in different places or set in different parts of the world, can contribute to an understanding of economically developing countries. Books specifically addressing cultural or religious aspects will be useful. Teachers will need to adapt these materials for pupils with limited literacy skills by reading or retelling the contents of books and using photographs, drama, role play and symbol supported work. Wiegand (1993), Rowe and Routh (1992) and Routh and Rowe (1993) provide suggestions for appropriate stories for addressing geographical aspects of economically developing countries. The points made at the start of this section about the need to counteract stereotypes is particularly pertinent in the selection of books.

Physical Geography

In an earlier chapter, reports were referred to which suggested that this area of geography is under emphasised except for work on the weather. Wiegand (1993) suggests some reasons why physical geography may be hard to understand, including the distinction between 'form as it exists at a given time' and 'the process of formation'. Further difficulties include the lack of precision in definitions used in physical geography such as 'hill', the multiple meaning of words such as 'range' and the variations in terminology for the same concept, in particular regional variations. The choice of language is particularly important with pupils with learning difficulties if their geographical

Figure 19 Adrian's record of the weather forecast, from George Hastwell School, Cumbria

understanding is to be enhanced. This applies to specific terms and to the wording of questions, as illustrated by my attempts during the session on use of buildings in the locality, described earlier in this chapter. While trying to get the pupils to identify a garage on the video I asked a class 'What do you do when the car breaks down?' The knowledgeable and accurate reply was 'Call the AA'!

The area of physical geography includes the weather and climate, rivers, landforms and vegetation, and soil. It lends itself to plenty of practical work and field work but remains conceptually rather inaccessible, perhaps because of all

AT/SoA	Content questions	Suggested activities
	What do I already know about this place? What will I expect to find? What will it be like? What do I think about it?	**Speculating** Drawing on pupils' previous knowledge Group discussions of perceptions culled from indirect experiences Imaginative work
AT3/3c AT3/4 AT3/6b AT3/6b AT3/6f	**Where is it?** What does it look like? Are there many or few people? What is it like to live there? What is the weather like? Are there many visitors? What is distinctive about it? What is the vegetation like? What animals live there? What is an ecosystem?	**Describing** Locating, describing, collecting, sorting information Classifying, using books, maps, atlases, globe, artefacts, people Climate graphs Photograph analysis, satellite imagery
AT3 AT3/8a AT3/6b	**Why is it like it is?** Why is it where it is? How did it get like this? Why did it happen? How is life affected by the place? How have people used or modified the place? How does it link with other places? Why is it similar to/different from where we live?	**Explaining** Connection between factors, e.g. climate/vegetation Causes of convectional rain Relationship between people and the environment Connection between places, e.g. links within and beyond the rainforest Factors which explain, e.g. how an ecosystem works Bringing together evidence from a variety of sources Making comparisons, e.g. forested/deforested Reasoning, e.g. effects of deforestation on soils, plants and animals
AT3/7d	**How is it changing?** How might things change? With what impact? What decisions will be made? Who will decide? Who will gain or lose? What are the alternatives? Will the changes bring improvement? For whom?	**Predicting** Using newspapers, magazines, television programmes and videos to examine the impact of deforestation and ranching on the rainforest Identifying the main groups affected and speculating on their future Simulation, role-play, computers, prediction models Examining global effects of deforestation
	What would it feel like to be here? What are the views of the people who live here? What do others think and feel about it? What do I think and feel about it? What can I do?	**Responding** Imaginative work Use of literature and television Diaries and photographs Making posters Finding out about organisations and pressure groups

Figure 20   The Brazilian rainforest – an enquiry approach (National Curriculum Council, 1993b, p.9, reproduced by permission of the School Curriculum and Assessment Authority)

areas in geography it relates least closely to people. The weather is more easily linked to people through activities such as matching clothes to weather, keeping weather charts using objects, photographs, clothes, symbols or drawings. In the example in figure 19, Adrian has recorded the weather forecast that he previously watched on 'Farmer's Forecast' and noted each day whether or not it had been correct. Each pupil in the group also had a map of the UK and recorded with a weather symbol what the weather would be like in each country. They then videotaped their own weather forecasts using the maps and charts. The pupils demonstrated learning of weather symbols and knowledge of the countries in the UK.

At Windsor School in Clacton, Essex, two days for geography work were identified in order to encourage all staff to develop this area of the curriculum. Four classrooms were designated as the four seasons and an appropriate environment and range of activities were set up in each. For example, the winter room was cold, displaying winter on all walls and a large pile of wool clothing to choose from when entering the room. In this room activities included constructing 'snow scenes' and drinking hot soup. Pupils spent one full day moving from room to room until they had experienced the four seasons. The

---

Soil and rock collections

Children can collect samples of the natural environment eg. soil, grass, leaves, rocks from around the school site. These can be put on a map or a model of the school to show where they come from or used to make model gardens.

Sorting and classifying activities

Children can explore soil by making models, balls and sausages with different soil types eg. sand, earth, shingle, clay. Use feely boxes to explore texture. Rocks can be sorted by colour, hardness, texture, etc.

Gardening activities

These allow the children to make the link with the living environment and explore plant growth. Collections of living creatures show how the soil is a home for them eg. wormery.

Puddle maps

Chalk outlines can be drawn around puddles in the school playground. Pupils observe them as they grow and shrink in wet and dry weather. Discussion about why they form and where the water drains are to be found around the school eg. gutter downpipes, gulleys and drains.

Miniature landscapes

Make model landscapes in the sand or earth tray or outside. Put in features such as hills, valleys, animals, roads and houses. Children can use toy equipment to quarry and build roads. Experiment with the effects of watering can 'rain' on the landscape. Indoor landscapes can be made using PE equipment.

Figure 21   Activities on landscapes, soil and vegetation (adapted from *Meeting Your Needs: National Curriculum History and Geography in Special Schools* (SLD/PMLD), Devon County Council, 1993, p.33-34)

second day was used to recall the activities that had taken place, a process enhanced by the use of a video recording of the previous day. This enabled further recording of pupils' responses to take place.

Activities about the rainforest can cover extensive areas of geography work since they involve studying a contrasting locality, physical features, settlement, economic and environmental issues, and the use of geographical skills. The example in figure 20 shows how an enquiry approach can be used specifically to address physical geography in this context although overlap with other areas is apparent. This example uses a wide range of teaching methods such as information retrieval from books, maps, atlases, globes, photographs, newspapers, television programmes, videos, simulation, role-play, computers and diaries, making it particularly useful for diverse groups of pupils.

The study of soil and vegetation will link closely with work in science, as do many other activities in physical geography. The list of activities in figure 21 for pupils with severe and profound and multiple learning difficulties has been adapted from the Devon County Council (1993) resources.

Physical geography can be approached through stories and poems such as those focusing on the weather, seasons, dramatic events such as floods, hurricanes and storms. Many of the well-known stories for young children such as *Winnie the Pooh, Postman Pat, The Wind in the Willows* and *Animals of Farthing Wood* include numerous references to the weather and its effects, the terrain, vegetation and animal habitats.

Human Geography

Human geography includes population, settlement, communication and movement. It is a relatively accessible area of geography for pupils with learning difficulties as all content such as housing, jobs and transport are concepts most pupils experience in some form, directly themselves or through those around them. At a more advanced level, we can expect pupils to begin to unpack the social, political and economic influences on the patterns of settlements, journeys, employment and land use. At a simpler level, pupils may begin to identify the similarities between these across different areas and the basic relationships involved, for example, identifying which shop sells which goods. This area of geography offers opportunities to develop economic and industrial understanding.

Settlement

Most of the published resources mentioned in this chapter cover aspects of settlement. In the *Our Town Activity Book* there are a number of activities relating to houses and homes, shops, and different types and uses of buildings. Children are encouraged to conduct a survey of the use of local shops by the people in their family, record their results and note items that can be bought in each shop, as shown in figure 22.

As part of a project on dwellings, pupils at George Hastwell School in Cumbria opened their own estate agents and compiled details of their own houses to display in their 'shop window'. The pupils discussed and described on

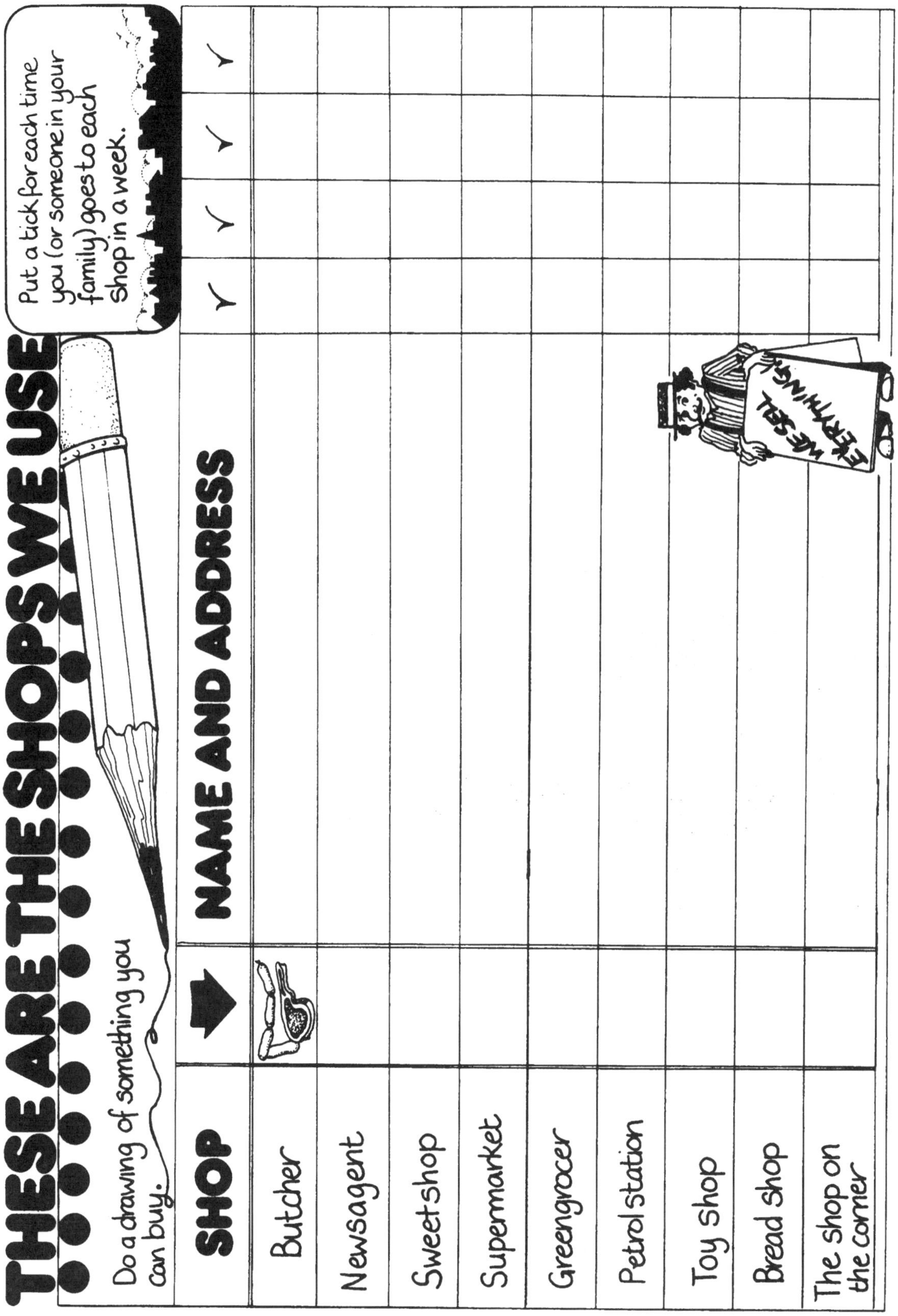

THESE ARE THE SHOPS WE USE

Do a drawing of something you can buy.

Put a tick for each time you (or someone in your family) goes to each shop in a week.

SHOP		NAME AND ADDRESS			
Butcher			✓	✓✓	✓
Newsagent					
Sweet shop					
Supermarket					
Greengrocer					
Petrol station					
Toy shop					
Bread shop					
The shop on the corner					

WE SELL EVERYTHING

Figure 22   A survey of shops we use from *Our Town Activity Book* (Hawkin, 1982)

Paul

For sale.

New semi detached house

on the Holbeck estate close to

the Abbey. Good country views.

Gardens to the front and back.

Big lounge, big dining room and big kitchen with lots of fitted

cupboards and a sink. Stairs leading to three bedrooms, two

big and one little. Fitted bathroom with white suite.

The house has central heating and a telephone point.

Offers over £60,000.

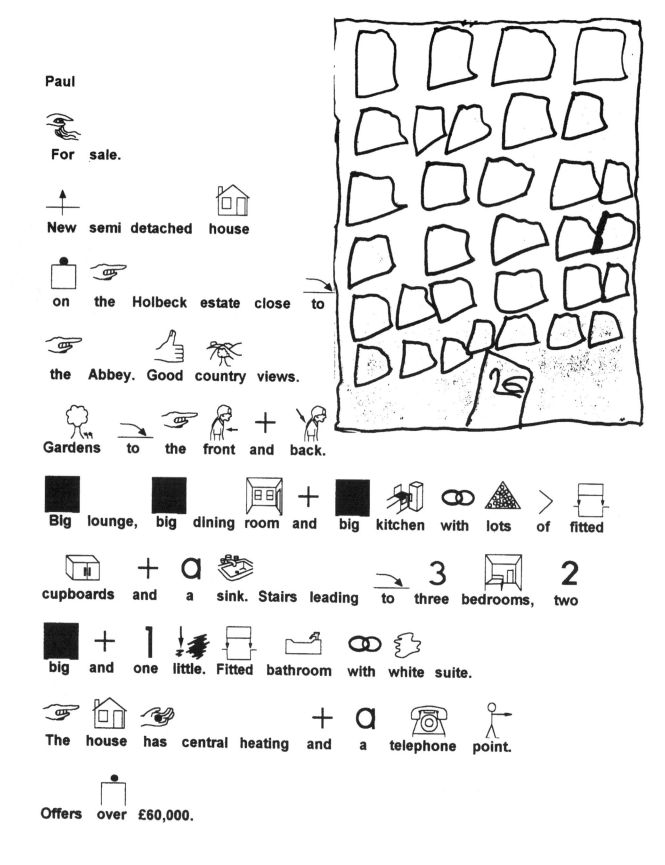

Figure 23 Paul's house details, from George Hastwell School, Cumbria

their details rooms in the house, position (which they located on a map), nearby amenities and type of building as shown in figure 23.

Some of the software programs reviewed in the next chapter support work on settlement.

Work

In *History for All* (Sebba, 1994) I described some activities designed to develop skills of interpretation of evidence. One of these activities involves identifying a person in school from a range of the items they use in their work. This activity is relevant to human geography which requires pupils to consider how goods and services are provided. Other examples of this are described in the next section on environmental geography. In the schools contributing to this work a range of activities has been undertaken. Pupils have been involved in sorting out items used by different people who work in the school, such as the caretaker, secretary, nurse and cook, and allocating the items to them using a photograph of them. At the end of the activity the pupils have taken the items back to the people involved to check with them that they belong to them.

In some schools this has led to extension activities such as interviewing people who work in or around school (such as the people delivering milk and post) about their work. These interviews have sometimes been tape recorded, or books have been made about the people describing their role. Commercially published versions of this type of material exist in series addressing 'people who help us' – but they often perpetuate stereotypes particularly in relation to gender. One exception of course is *Mrs Plug the Plumber* (Ahlberg and Wright, 1980), aimed at young children, although most of the other books in this 'Happy Families' series conform to traditional gender stereotypes. In one school the activities on work roles led to arrangements being made for pupils to do short spells of 'work experience' with different people who work in school.

Journeys

Many of the resources described elsewhere in this chapter include activities addressing journeys, in particular surveys of how pupils get to school, traffic surveys and comparative transport between the local area and a contrasting area. At George Hastwell School they undertook a project on the use of one particular road which included traffic surveys, questionnaire design, writing to local firms to find out what cargoes they transport, questioning drivers about their routes and loads, and mapping the route taken by the road. One pupil produced his work on this project as part of a City and Guilds Foundation Programme, while another, Sarah, provided the completed table in figure 24 to show the replies received from drivers about their loads and destinations.

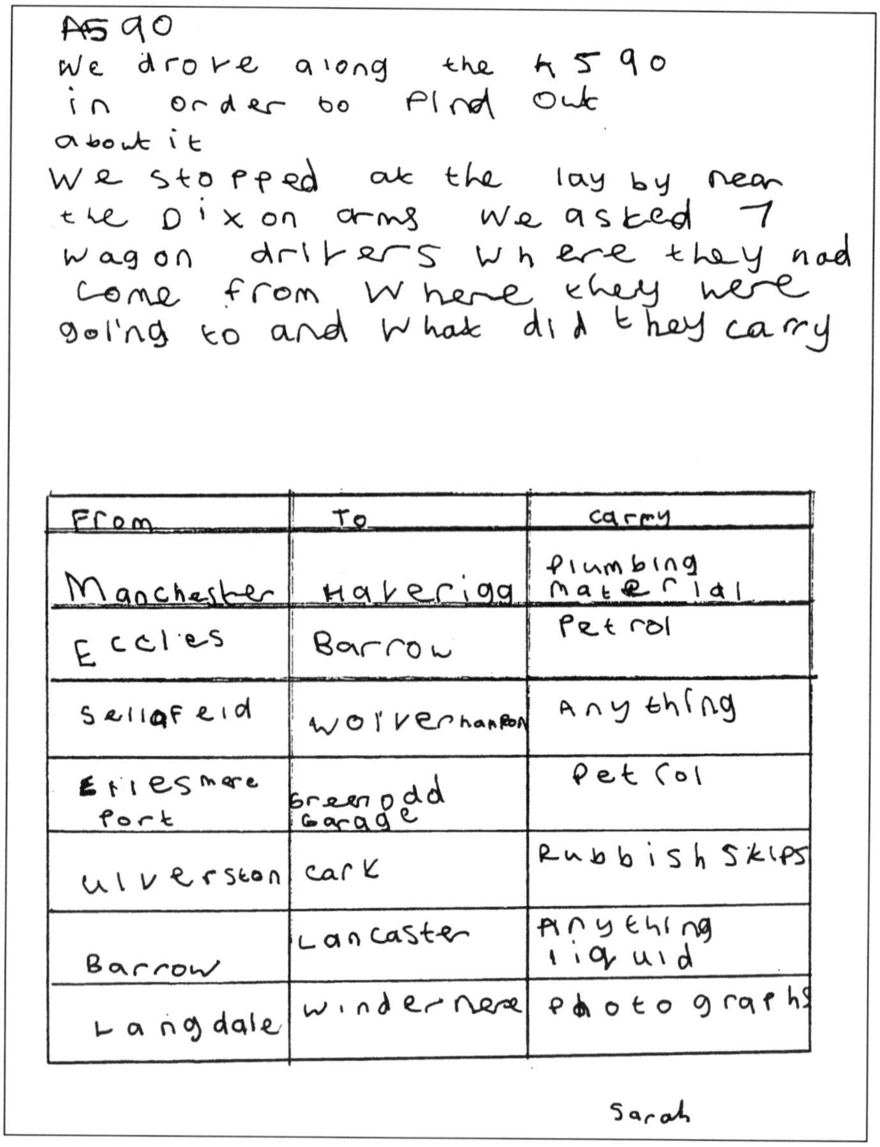

```
A590
We drove along the A590
in order to find out
about it
We stopped at the lay by near
the Dixon arms We asked 7
wagon drivers Where they had
come from Where they were
going to and what did they carry
```

From	To	carry
Manchester	Haverigg	Plumbing material
Eccles	Barrow	Petrol
Sellafeild	Wolverhampton	Anything
Ellesmere Port	Greenodd Garage	Petrol
Ulverston	cark	Rubbish Skips
Barrow	Lancaster	Anything liquid
Langdale	windermere	photographs

Sarah

Figure 24 Sarah's information on routes and loads from interviews with drivers, from George Hastwell School, Cumbria

Environmental Geography

Environmental geography is about the effect that human behaviour can have on the environment, management of the environment, sustaining and improving the quality of the environment and preferences about the environment. The proposed revision of the curricular requirements for 1995 have removed the study of where common materials are obtained from and how they are extracted because these are addressed in the proposed science curriculum, but it remains relevant to the provision of goods and services specified within human geography and economic activities in key stage 2. Examples of this area of work are included in this section as they are seen as relevant geography whether or not they remain specified curricular requirements.

The role of values in environmental geography is a sensitive area. Decisions affecting the environment are made by people, usually on the basis of values and sometimes influenced by vested interests. In addition, people hold different values about the aspects of the environment that should be destroyed, conserved or developed. Hence, teaching environmental geography involves teaching pupils about the values of different people, but the point at which this teaching is considered to become indoctrination of the values themselves is difficult to judge. Wiegand (1993) suggests that teachers must teach pupils that the study of people's interrelationship with the environment is necessary and worthwhile and that pupils should be interested in, and concerned about, the environment.

Natural resources

At John Smeaton Community High School in Leeds, pupils with severe learning difficulties participate in their year group's geography lessons. Jamie, a year 8 pupil, dictated his replies to the questions on the worksheet to a support teacher who wrote them on the sheet for him to copywrite underneath as shown in figure 25. The sheet comes from the *Key Geography* materials (Waugh and Bushell, 1992). This example demonstrates how the lesson gave Jamie the opportunity to learn some geography while targeting a writing priority.

Activities which involve tracing an everyday commodity back to its source can create an effective opportunity for learning. The example in figure 26 from the East Sussex County Council (1991) schools for pupils with severe learning difficulties provided relevant work in other curricular areas as well as addressing many areas of geography. The participation of pupils interviewing the farmer and milkman, milking a goat and investigating what you can make from milk, makes these activities appropriate for diverse groups of pupils, many of whom will have limited or no literacy skills.

Managing the environment

A recycling project was observed in a special class attached to a large secondary school. The pupils brought in a variety of rubbish from home and were introduced to categories of waste and methods of waste disposal. This was related to protecting the environment through appropriate disposal of litter and collection of items for recycling.

The pupils collected the rubbish from each classroom in the secondary school at the end of a day and labelled it. They constructed a simple bar chart to indicate the amount of rubbish and type of rubbish that came from each class. A simple map of the school and the immediate vicinity around it was used to note the sites of litter bins, the shops and the areas in which litter was greatest. They conducted interviews with the headteacher and some of the secondary school pupils to find out what type of rubbish they throw away, where most of it comes from (eg. local sweet shop) and their views about the rubbish around school and about recycling. The interviews were taped and used as a basis for further discussion.

As a result of this project, additional litter bins were introduced at sites identified as problem areas by this group of pupils. Subsequent monitoring of the rubbish by the pupils demonstrated the positive effects of this strategy. Related activities undertaken included making posters about litter to display in

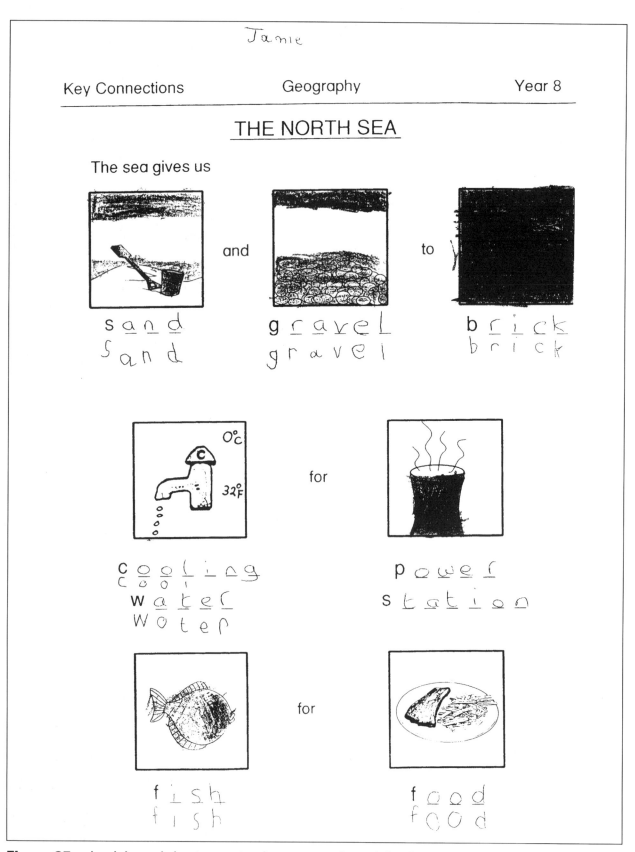

Figure 25 Jamie's worksheet on natural resources, from John Smeaton Community High School, Leeds

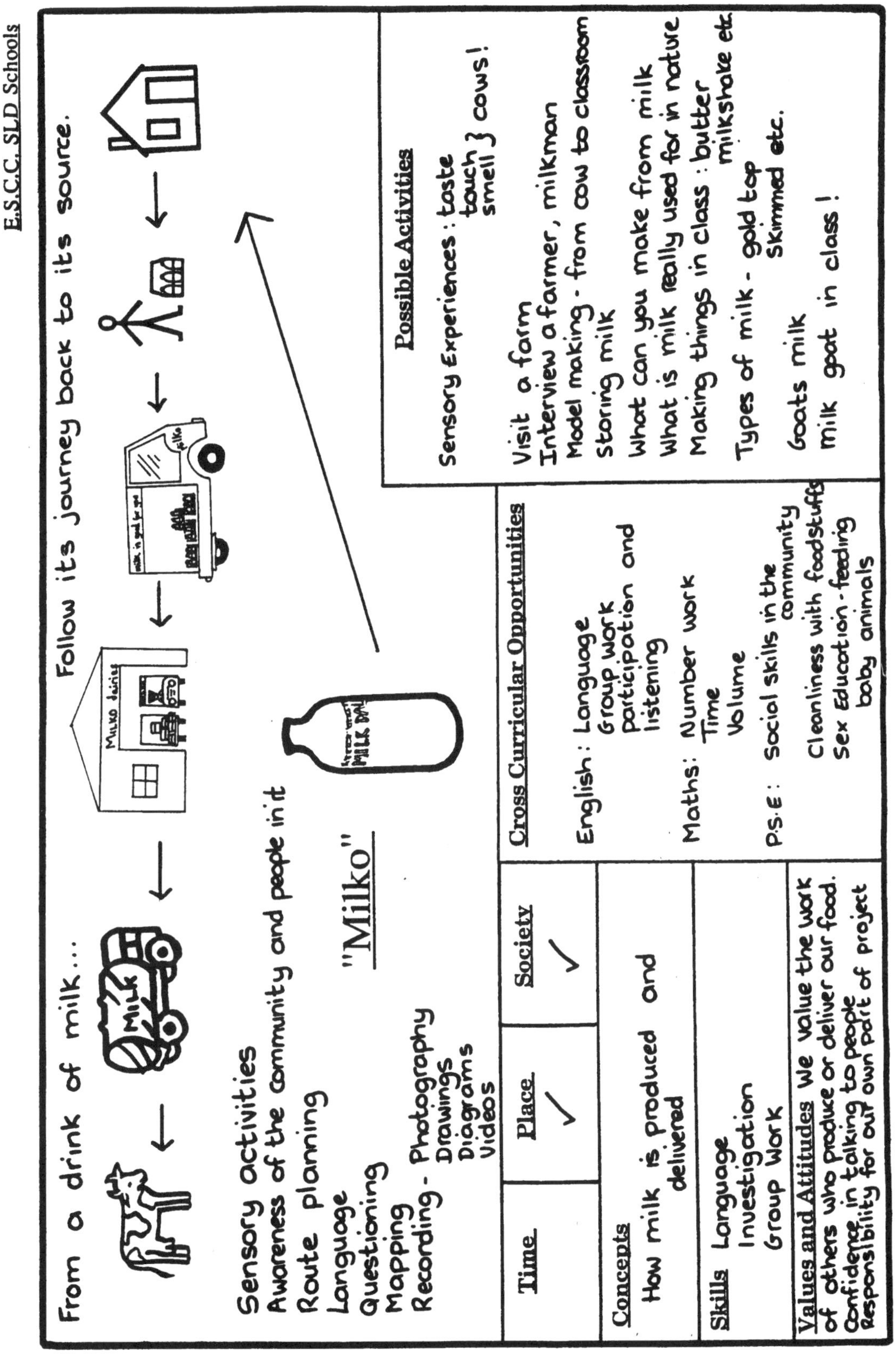

Figure 26 'MILKO' : following milk back to its source (East Sussex County Council, 1991, Activity Sheet 16)

the school, drama about the environment, visits to the supermarket to find out what they do with their waste, to Friends of the Earth and to the local waste disposal plant. The computer was used to access activities for some pupils in order to make charts, write letters and record results. Self-recording of parts of the project contributed to pupils' records of achievement.

The teaching approach was exemplary in a number of aspects. The purpose of each activity was explained to the pupils at the start of each session. Active involvement for every pupil was enabled through the range of tasks, variation in teaching approaches and by appropriate use of a classroom assistant. Pupils were able to celebrate their achievements through their own display of work to which they regularly added items. Activities were systematically recorded by teacher and pupils.

Environmental issues

A simulation devised by the Suffolk County Council Humanities Advisory Team (1989) in order to promote economic understanding involved a proposal to build a brewery on the site of an old school in the village of 'Pontafon'. The pupils used the concept keyboard to build up a map of the village of which they then constructed a model. They retrieved information about the residents of the village using the concept keyboard map. Each child was allocated the role of one of the residents, local traders and others representing various interests. A public meeting was held with the teacher in-role as a planning officer. The campaigns on behalf of the different interest groups involved were then developed using posters, debate, canvassing and written submissions. All the pupils in a group with diverse needs in a junior school were able to contribute to this activity and the interviews with pupils at the end of the project revealed that understanding about environmental issues had increased.

Similar environmental issues are addressed through the Site Seeker software program (Northwest SEMERC) reviewed in the next chapter. In addition, Peter (1994) describes examples of using drama to explore environmental issues. Rowe and Routh (1992) and Routh and Rowe (1993) suggest many suitable stories which address pollution, ecology, conservation and environmental conflicts. For young children, the well-written stories of *The Animals of Farthing Wood* (by Colin Dann) manage to encompass a wide range of these issues within an exciting and enchanting context. My favourite is *Dear Greenpeace* (James, 1991), a series of letters exchanged between little Emily and Greenpeace, from which the following has been extracted:

Dear Greenpeace,

I love whales very much and I think I saw one in my pond today. Please
· send me some information on whales, as I think it might be hurt.

Love Emily

Dear Emily,

Here are some details about whales. I don't think you'll find it was a whale you saw, because whales don't live in ponds, but in salt water.

Yours sincerely, Greenpeace

In this chapter, activities addressing a wide range of content in geography have been reviewed. The examples chosen are particularly accessible to all pupils and make few assumptions about their current skills. The next chapter looks specifically at the ways in which information technology and group work can be used to increase access in geography.

CHAPTER 4

Specific Teaching Strategies to Increase Access to Geography

Introduction

Access to geography activities can be increased in many different ways. This chapter focuses on only two of these: group work and information technology. Access to geography should not be limited because of the reading or writing demands involved in the task. Meeting individual needs by adapting teaching approaches is sometimes reduced to modifying worksheets. While this approach has a place, it is by no means the most effective method of ensuring that diverse needs are met.

Procedures for adapting and simplifying activities in geography that require reading and writing are similar to those used in other subjects. Examples of work in geography for pupils with learning difficulties in the secondary school, including the use of cloze procedure, underlining, prediction, sequencing, highlighting and extracting and reorganising information, are provided in Clarke and Wrigley (1988). They also demonstrate the considerable variation in the reading demands generated through different texts in geography and the inconsistencies between readability scores given for the same text (a history example) using different tests.

It is important to identify clearly from the outset the geographical understanding that it is intended to develop through any procedure adopted, and to avoid the trap of pupils with learning difficulties wasting time colouring or undertaking simplistic tasks while others pursue geography. The rest of this chapter focuses on two areas considered to make a particularly strong contribution to accessing geography to a greater number and range of pupils. The first approach to be described is group work.

Increasing Access Through Group Work

The value of rigorously planned and sensitively implemented group work for increasing access to the curriculum for a wide range of pupils has been discussed fully elsewhere (for example, Sebba *et al,* 1993). In that publication we reviewed studies demonstrating that merely seating pupils in groups is unlikely to have much effect on their learning although well structured group work can contribute to enhanced learning, especially for pupils with learning difficulties. It is important not to become a 'group work groupie' but to recognise that different methods of classroom organisation are suitable for different content and activities. In geography, work addressing environmental issues and geographical enquiries lend themselves better to group work than topics which are predominantly knowledge based.

Careful planning and organisation are needed in order for group work to lead to enhancement of geographical skills. Effective group work enhances the value of discussion in developing essential skills in geography, including expressing a point of view and listening and responding to the ideas of others. These skills are, in the terms of the OFSTED (1994) inspection criteria, learning skills which contribute to the quality of learning across the curriculum.

However, many pupils with learning difficulties experience particular difficulties with communication which may belie their levels of understanding. It is therefore important to consider the organisation of group work that may best enable them to communicate. For some pupils with severe or emotional and behavioural difficulties, starting with work in pairs might be more realistic, moving on to small groups when tolerance and awareness of others have developed.

In Sebba *et al* (1993) we provide examples of methods of organising groups, including 'jigsawing', which might enhance the contribution of all group members. The allocation of roles, responsibilities and materials and careful consideration of the task and timing can encourage involvement. Tape recorders used with the pupils' consent can provide a basis for pupils and teachers to evaluate the session, although if some pupils are using symbols or signing for communication, a video will be more helpful.

Many of the examples referred to earlier in this book utilise group work techniques. The use of simulations such as *Pontafon* or software such as Site Seeker, reviewed later in this chapter, enable pupils to consider different perspectives on the same issues and can use group work in which each group adopts a different perspective. Another method is to give each group a different part of the task and to ask them to bring their work together when the task is complete in order to build up a total picture. For example, an enquiry about a specific location could be tackled by splitting the location up into areas where different groups of pupils collect different data as part of a whole class project. If each pupil individually, or in the whole class, had to address every aspect, it is likely that only superficial coverage could be achieved. Furthermore, if flexibility is provided on the form of presentation the groups use to report their findings, access to the activity for those pupils whose literacy skills are limited will be increased. These are whole class activities designed to involve children in their own learning and to place a responsibility on each pupil for the quality of the learning of the whole group.

An example of group work which illustrates these points comes from the Wren Spinney School (undated) Module on learning about maps. Some pupils working in twos or threes conduct a survey of where people live by asking the other pupils in school, providing a limited range of options. Some pupils are responsible for collating information related to a specific place. Another group of pupils is responsible for entering it into a database in order to produce block graphs or pie charts. As each group gathers its information one pupil, the 'envoy', is sent to report its findings to the group collating the information. This group enters the data on their charts and sends a further envoy to the group entering information on the database. The envoying technique encourages individual responsibility within the group, reporting skills, and enables information to be shared.

Increasing Access Through Information Technology

The use of computers in teaching is viewed cynically by many teachers. Lack of spontaneous access to machines, insufficient staff development and the availability of plenty of software that is unhelpful, have all contributed to its negative image in some schools. In addition, there is a temptation to overuse the computer with pupils whose behaviour or learning difficulties appear to subside when they are working on it, particularly in a primary or special school classroom where one machine may be permanently available. Having dispensed with the warnings, the potential for developing geographical skills through the use of appropriate software is considerable, as this chapter will attempt to demonstrate; but as OFSTED (1993b, p.5) note, it is under utilised at present:

> Although a good range of information technology facilities was available in most schools, few geography departments made use of this provision to enhance the geography teaching in Key Stage 3.

Computers may be used to motivate pupils to explore geographical concepts where other methods lack appeal. This may involve simulations, adventure games, constructing maps, databases or information retrieval. The Curriculum Council for Wales (1991) provide a detailed analysis of ways in which information technology can contribute to geography.

It is worth considering why the computer should be used rather than other available methods, particularly where the reading demands in the software are comparable to those in the written resources. Word processing packages and databases have the potential to enable the research skills and communication of geographical ideas to be enhanced. There is a vast range of software available and the review here is limited to programs that illustrate specific points relating to access for pupils with learning difficulties.

Adventure games

Adventure games can develop geographical understanding almost without the pupils being aware that they are learning. The choice of games should be guided by the criteria on the software evaluation sheet in figure 33 below. It is particularly important to consider the geographical skills and understanding that the program covers and the access for pupils whose reading skills are limited. Many of the games are rather long for classroom use so the flexibility to stop at any point in the adventure and return to that point at a later time is also important.

There are a few software companies that demonstrate a definite commitment to pupils with learning difficulties through careful consideration of accessing issues in software design. Sherston Software (address in the Resources list) is one of these. Their graphics and sound effects are particularly impressive. The more recent programs, such as the adventure ones reviewed here, use simple, limited vocabulary which appears on the screen in small amounts, and pupils who cannot read soon learn to click the mouse over the instruction on the bottom right of the screen to move on even if they are unable to read it. The programs are written so that they can be set up at various points in the story, enabling pupils to work over several shorter sessions. At the time of writing (1994), in the area of geography Sherston offer three adventure games (Nature Park Adventure, Crystal Rain Forest and Badger Trails), one mapping and adventure program (Mapventure), and several programs designed to explore environments (Viewpoints, Peek-a-boo around our Town, and Peek-a-boo around our House). Many other programs that they produce include activities relevant to work in geography.

Nature Park Adventure

Nature Park Adventure (Sherston Software) invites the pupils to rescue the rare butterflies that have disappeared from the Nature Park. As with many Sherston adventure programs, there are the usual appropriate enemies, in this case a trogg and two glob monsters, who have taken the butterflies to sell them. The support materials include work cards, fact cards, nature cards and some related mathematics activities. The graphics are excellent and the language simple and clear. The program covers geographical skills (directions, mapping), physical geography (features, habitats), environmental geography (conservation, environmental issues) and some references to human geography.

Crystal Rain Forest

The Crystal Rain Forest (Sherston Software) is an adventure game about the planet of Oglo, where the last remaining rain forest is under threat from the Cut and Run Gang. The pupils must save it by solving a series of problems involving Logo. A simplified version of Logo is included in the pack. The graphics and sound effects make the adventure stimulating and rewarding to use. The geography content includes conservation, effects of climate and mapping skills.

Badger Trails

Badger Trails (Sherston Software) includes a videotape and support materials in addition to the software. The story book included in the pack tells of a badger who disappears from a set the children are observing and needs to be helped to find its way back home. The program simulates the badger's sensory experiences and the video provides a record of the real discovery of a badger set with an opportunity to observe the badgers. The materials cover mapping skills, physical and environmental geography.

Mapventure

Part of The Mapventure program (Sherston Software) attempts to develop skills

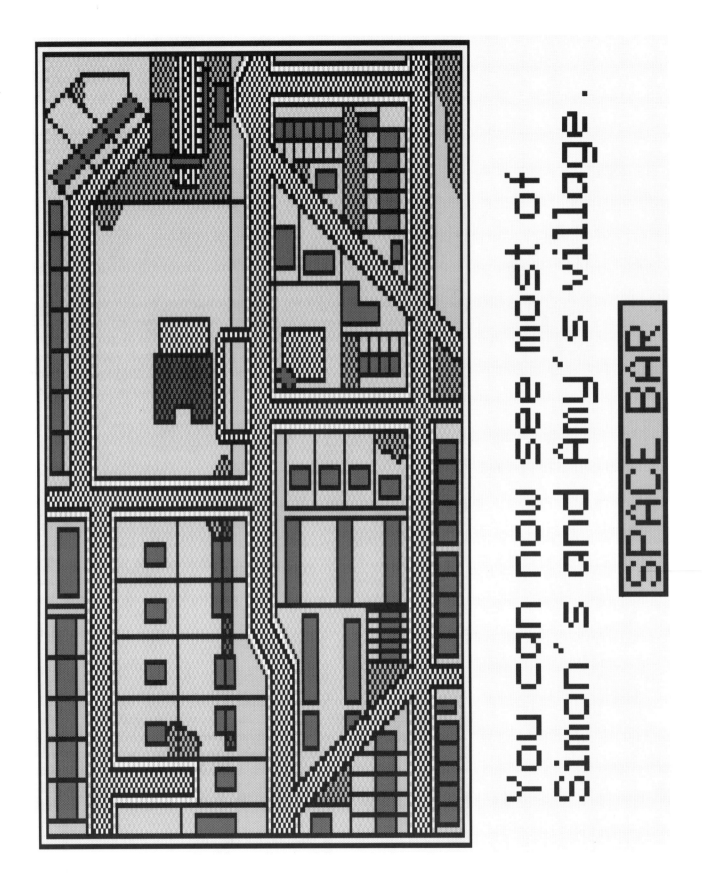

You can now see most of Simon's and Amy's village again.

SPACE BAR

Figure 27 A screen from Mapventure (Sherston Software)

in using plans and maps. The other part provides an adventure in which the pupils must find their way and avoid being trapped by the Red Devil Gang by keeping away from areas defined by stated co-ordinates. The graphics are motivating and the activities more accessible than traditional textbooks. The program covers plans, maps, contours, symbols, co-ordinates and directional skills including compass points. The section on contours is particularly helpful as it enables pupils to view the way in which contours are constructed through a series of moving screens. The program addresses predominantly geographical skills. A sample screen is shown in figure 27.

Viewpoints

The Viewpoints program (Sherston Software) offers the chance to explore environments by zooming in on specific features to examine them more closely. Anything which is discovered can be photographed, which enables it to be transferred to the database. The graphics and sounds on the animations are captivating. In order to move around the environment directional instructions must be given and the screen enables several views from the same point to be seen simultaneously. The program covers geographical skills such as directions and mapping, physical geography, in particular, animals and vegetation and environmental geography. A sample screen is shown in figure 28.

Peek-a-boo Around our House / Peek-a-boo Around our Town

These two Sherston Software programs enable exploration of a house and town at a most basic level. Pupils indicate with arrows (which can be operated by a single switch) where they would like to look inside. The rooms in the house can be explored to see which objects are to be found in which room. The shops in the town can be explored in order to find out which goods are sold by which shop. The example in figure 29 illustrates the chemist and post office from 'around our town' which can then be entered to find out what goes on inside each and what goods or services they provide. The programs cover directional and mapping skills and human geography.

Blob, Going Places and GridIT

Widgit Software (address in Resources list) demonstrate a commitment to imaginative programs for pupils with learning difficulties and have led developments in the use of symbols systems with the computer. Most of their programs are developed in conjunction with schools or services who trial and evaluate them before they are made more generally available. The widely used Blob and Going Places programs and the more recent GridIT program all include mazes at different levels which are appropriate for work on geographical skills. GridIT enables pupils to plan and program mazes at varying levels of complexity, encouraging developments in mathematical and geographical skills. Two screens showing mazes at different levels from GirdIT are shown in figure 30.

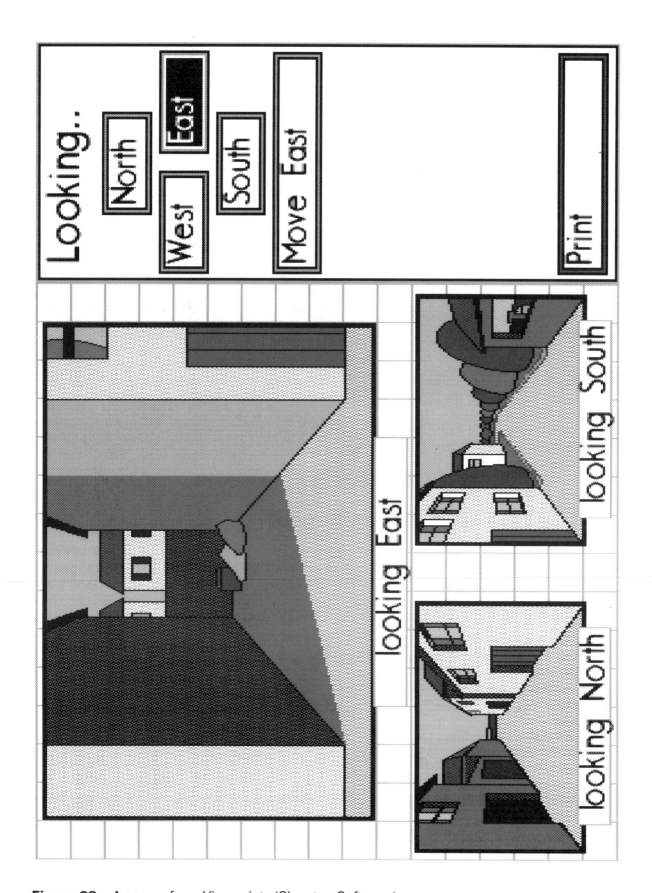

Figure 28 A screen from Viewpoints (Sherston Software)

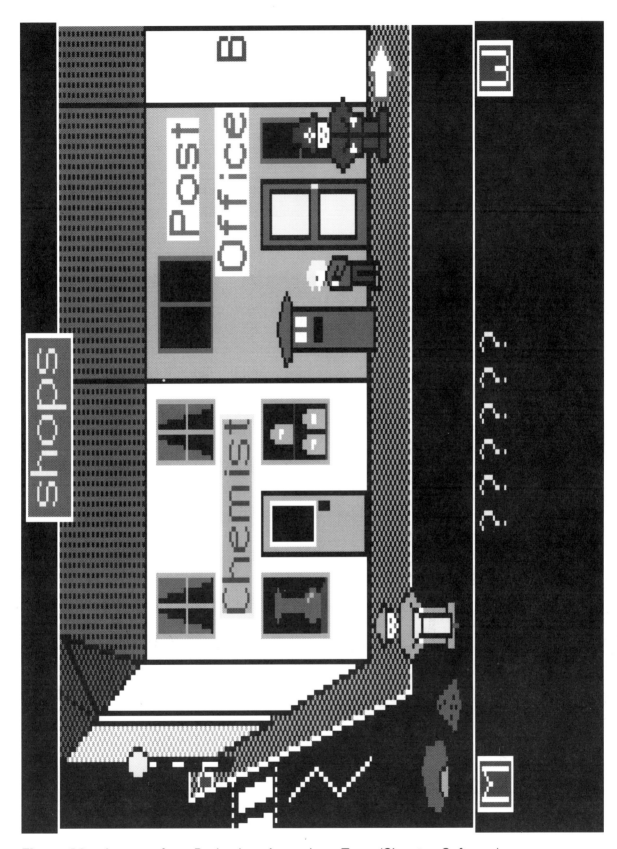

Figure 29 A screen from Peek-a-boo Around our Town (Sherston Software)

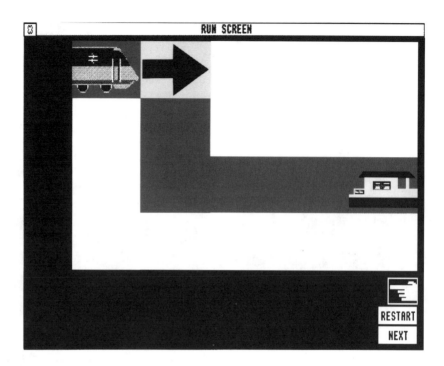

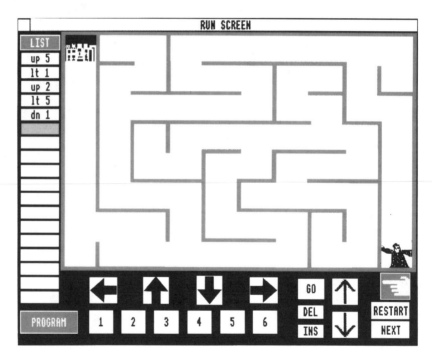

Figure 30 Two screens showing mazes at different levels from GridIT
(Widgit Software)

Site Seeker

Site Seeker (Northwest SEMERC) is a National Council for Educational
Technology 'Blue File' (freely copiable for educational use) program distributed
by Northwest SEMERC (address in resources list). Site Seeker simulates the

decision-making process involved in choosing a site for a development. It is a content free program in the sense that it enables you to create your own files, for example, in order to address an issue of local relevance. However, it includes an example which involves choosing a site for a new inner city playground, with five possible sites from which to select.

The program offers the choice of viewing a map of the area, close up views of each site, more information about the site including size, terrain, cost, the surrounding area, population, crime and road safety. When information on these is provided pupils are asked to decide whether this is an advantage, disadvantage or neither, and their responses are recorded. Similarly, they can seek the opinions of children, police officers, old people, councillors, community workers, shopkeepers and parents about each site and when receiving their comments pupils make decisions about whether these are advantages, disadvantages or neither, which are recorded. This process enables them to consider the evidence for and against a site at any time during the exercise and to print out this information. The support materials include letters, notices and other written sources relevant to the playground issue. This program is ideal for tackling a local planning issue and lends itself very well to group work.

My World/Geography Key Stage 1

My World/Geography Key Stage 1 (Northwest SEMERC) runs off the 'My World' program, which at the time of writing is only available for the Archimedes. The program enables the creation of streets and towns, exploration of siting and use of buildings, creation of aerial views on buildings and towns, and an activity to label the countries, cities, mountains and rivers of the UK. Other activities include matching goods to the shops which provide them, following directions and recording the weather. Overall, the suite of programs contributes to most areas of geography at the early levels.

Urban Studies

Urban Studies (Science Education Software, address in Resources list) can be used with pupils working at advanced levels, as it covers data-processing aspects of geography including demographic data from cities, traffic flow, use of questionnaires about the environment, housing surveys, commodities and services. Many of these activities require a high level of literacy and mathematical understanding. However, the first part of the program enables graphical street profiles to be generated by specifying use of building, size and number of storeys. This enables pupils with learning difficulties to construct a profile of a local street as part of a study, as shown in figure 31. This part of the program addresses human geography and local place studies, although the suite of programs overall addresses many other areas of geographical work.

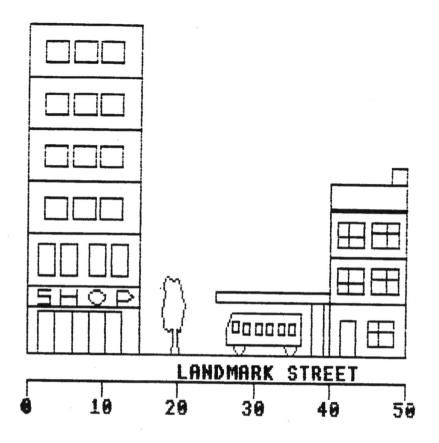

LANDMARK STREET

0   10   20   30   40   50

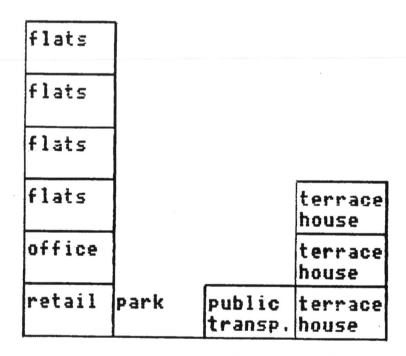

Figure 31   A street constructed from 'Graphic Street Profile, Urban Studies' (Science Education Software)

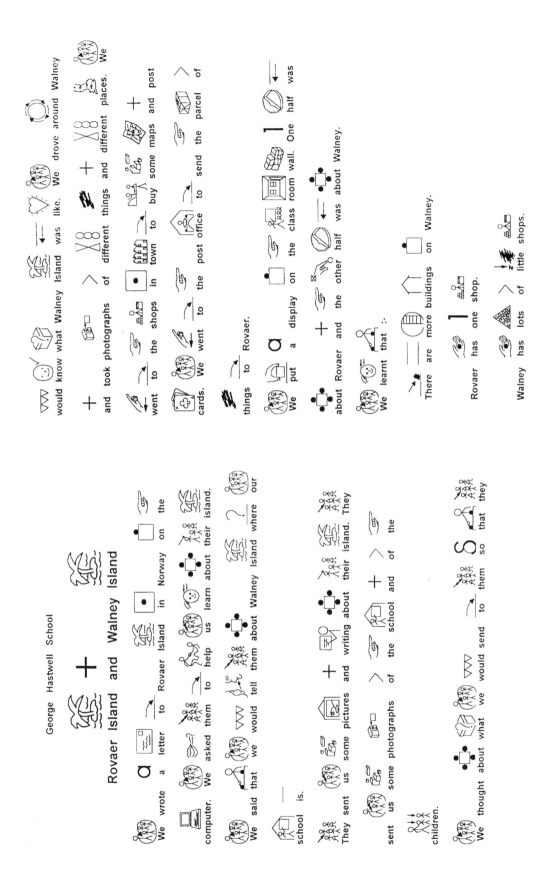

Figure 32 Part of an account of a school link using symbols, from George Hastwell School, Cumbria

69

Other uses of computers to increase access in geography

Many databases, spreadsheets and other content free programs are useful resources for work in geography. Software atlases are available which usually include databases with information about different countries that encourage information retrieval, although these often require high reading levels. Producing graphs, charts and simple tables enhanced through the use of graphics can make geography more accessible to pupils with learning difficulties when carefully chosen. Some undoubtedly put off more pupils than they help but the quality of presentation is improving all the time.

The use of computer generated symbol systems for recording and communicating about work in geography provides further access for pupils with limited literacy. Widgit Software has several programs including From Pictures to Words, Using Symbols and more recently The Writing Set, which enable pupils to communicate using computer-generated 'Rebus' symbols. The example in figure 32 shows part of George Hastwell School's account of their link with the school in Norway, described more fully in the last chapter.

Evaluating software

It is helpful to have some clear criteria to refer to when considering purchasing software to use in geography teaching. The review sheet in figure 33 was designed to help do this and should be adapted to meet your own requirements.

Name of Program:

Supplier:

Type of program eg. adventure, word processor, database, etc.:

Suitability of program for the pupils you teach:

   Note reading requirements:

User friendly?:

Best use in small groups, individuals, etc.:

Geographical content covered:

Geographical skills covered:

In what ways might this program contribute to pupil's learning in geography?:

Cross-curricular content:

Figure 33 Software review sheet

In this chapter, two of the many approaches to increasing access to geography for *all* pupils have been explored. It is recognised that not all teachers will feel confident using information technology or group work. However, all are approaches that in most schools some teachers have used. This provides the possibility for teachers to extend their skills in a chosen area by team teaching, observing another teacher or exploring techniques through discussion, perhaps using practical examples from video, recording sheets or samples of work. These practices are the basis of effective staff development, one of the focuses of the final chapter.

CHAPTER 5

Concluding Thoughts

Principles for Increasing Access to Geography

Throughout this book certain principles are repeatedly illustrated. These are:

- seek relevance, for example, through focusing on people or homes;
- develop activities which take account of pupils' interests and experiences;
- take every opportunity to use geographical language across the curriculum, for example, at the simplest level, 'up', 'down', 'behind', 'in front of', 'near' and 'far';
- look for opportunities to address pupils' individual priorities within geographical activities, for example, communication skills or mobility within field work activities;
- use a variety of resources, for example, video, objects, photographs, maps, other people, buildings and sites;
- vary the teaching approaches as much as possible, using drama, information technology, and working individually, in pairs, small groups or whole class;
- share the purposes of sessions with the pupils;
- invite pupils to record information, responses and reflections, using drawings, photographs, symbols, computers or tape recorders, as appropriate;
- use total communication such as symbols, signing and Braille to increase pupils' access to activities.

Most of these principles will benefit *all* pupils, not just those with identified learning difficulties. The survey of history and geography in primary schools (DES, 1989) noted that the best geography work was achieved when activities were of a practical nature and based on pupils' interests and experiences. These principles relate closely to the criteria for quality of teaching and quality of learning used on school inspections (OFSTED, 1994), which were themselves developed partly from the literature on school effectiveness.

In what ways can geography contribute to the process of preventing or overcoming learning difficulties? The Curriculum Council for Wales (1991) suggests that geography can contribute to meeting diverse needs by providing a context that has an immediacy and is relevant, giving pupils opportunities to draw on their own experience and reflect on that of others, encouraging 'learning' skills such as those involved in geographical enquiry, and developing skills to operate safely in new environments. This book has provided examples of ways in which geography contributes to other subjects and to pupils' learning more generally.

Staff Development

Many of these principles are no different whichever subject is being accessed but geography has been underdeveloped in work with pupils with learning difficulties. This is partly because, as with other subjects, those teachers who find themselves delivering geography, particularly in primary and special schools, tend to be non-specialists. In chapter 2, it was suggested that part of the role of the geography co-ordinator is to support teachers whose confidence and knowledge of the subject is developing. The first step may be to ensure the co-ordinators themselves have sufficient staff development opportunities to ensure that they can undertake this role.

Staff development for non-specialists has been enhanced in England and Wales in recent years by the introduction of subject courses, designed to develop knowledge and understanding in the subject rather than focusing on teaching methodology. These courses have enabled teachers in primary, middle and special schools to develop their subject expertise and use this to establish a role in curriculum and staff development in the school. The evaluations of the courses in which I have been involved suggest that these intentions are largely met and the continuation of these courses is critical in ensuring that developments continue.

Some of the examples included in this book have come from work initiated by courses. However, for the schools to benefit from the course there needs to be a commitment through time and attitude to enabling the course participant to provide appropriate support to other teachers. This goes beyond merely encouraging them to produce schemes of work or lesson plans for others to adopt. It should involve team teaching, partnership schemes or 'coaching' to ensure developments in classroom practice can become established.

Monitoring and Evaluation

The first set of reports on school inspections of secondary and middle schools in England and Wales reflects some common issues. One of these is the lack of well established systems for internal monitoring and evaluation. While both monitoring and evaluation are built into the school development planning process, the rigour with which they are carried out is rather variable. The school should monitor many aspects such as curricular coverage, homework, marking, standards of work and the quality of teaching. In relation to geography, the subject co-ordinator will need to ensure curricular coverage is recorded through an auditing procedure (see National Curriculum Council, 1992, for examples of ways of doing this) and that other teachers are teaching the subject well. In chapter 2, approaches that can be used by the co-ordinator to keep track of the coverage in geography were suggested.

The monitoring procedure will provide information on what is happening which can then be evaluated by considering it against specific criteria. This might focus in geography on the progress of pupils in the areas described in chapter 1: geographical skills, places, physical, human and environmental geography. Evaluation of the use of resources in these areas will also be needed.

Why Geography for Pupils With Learning Difficulties?

This book attempts to entice the cynics who believe geography is limited in what it can offer pupils with learning difficulties to reconsider their position. It is also aimed at those teachers who are enthusiastic but are unsure where to start. The first step is to secure a commitment to the subject yourself.

Wiegand (1993, p.1-2) provides an appropriate conclusion for this volume:

> .....geography is necessary and worthwhile. After all, it deals with some of the most basic values of all – such as survival and the quality of life.

The opportunity for *every* pupil to develop these is in your hands.

Quote

References

Ahlberg, A. and Wright, J. (1980) *Mrs Plug the Plumber.* London: Penguin.

Brace, S. (1992) 'Project: a village in India', *Primary Geographer*, 10, 6-9.

Catling, S. (1984) 'Building less able children's map skills', *Remedial Education*, 19, (1), 21-27.

Catling, S. (1993, revision) *Mapstart 1, 2, & 3.* Glasgow: Collins Longman Atlases, Collins Educational.

Clarke, J. and Wrigley, K. (1988) *Humanities for All: Teaching Humanities in the Secondary School.* London: Cassell.

Clarke, J. (1992) 'Geography: another time, another place', *in*: K. Bovair, B. Carpenter and G. Upton (eds.) *Special Curricula Needs.* London: Fulton.

Cunliffe, J. (1983) *Postman Pat's Tractor Express.* London: Scholastic Book Services Inc.

Curriculum Council for Wales (1991) *Geography in the National Curriculum: Non-Statutory Guidance for Teachers.* Cardiff: Curriculum Council for Wales.

DES (1986) *Geography from 5 to 16: Curriculum Matters 7.* London: HMSO.

DES (1989) *Aspects of Primary Education: The Teaching and Learning of History and Geography.* London: HMSO.

Devon County Council (1993) *Meeting Your Needs: National Curriculum History and Geography in Special Schools.* Exeter: Devon County Council.

DfE (1994) *Code of Practice.* London: HMSO.

Doran, C. (1992) *National Curriculum Geography and Pupils with Severe Learning Difficulties.* Unpublished assignment for Advanced Diploma, University of Cambridge, Institute of Education.

East Sussex County Council (1991) *A Hitchhiker's Guide to Humanities.* Lewes: East Sussex County Council.

Elliott, D. (1994) *Life in Europe.* Hove: Wayland.

Gregg, Sister M. and Leinhardt, G. (1994) 'Mapping out geography: An example of epistemology and education', *Review of Educational Research*, 64 (2), 311-61.

Harrison, P. and Harrison, S. (1988) *Discover Maps with Ordnance Survey.* Southampton: Ordnance Survey/ Holmes McDougall.

Harrison, P. and Harrison, S. (1991) *Time and Place: History and Geography for Key Stage 1: Teachers' Source Books.* London: Simon & Schuster.

Hawkin, T. (1982) *Our Town Activity Book.* Cambridge: Cambridge University Press.

Humberside County Council (1992) *Access to Geography: Geography for Children with Special Educational Needs: Practical Guidelines Series.* Hull: Humberside County Council. Available from: Educational Publications Unit, Humberside Education Centre, Coronation Road North, Hull HU5 5RL.

James, S. (1991) *Dear Greenpeace.* London: Walker Books.

McNeill, C. and Renfrew, T. (1990) *Start Orienteering 1-6.* Doune: Harveys.

National Curriculum Council (1990a) *Curriculum Guidance 4: Education for Economic and Industrial Understanding.* York: National Curriculum Council.

National Curriculum Council (1990b) *Curriculum Guidance 5: Health Education.* York: National Curriculum Council.

National Curriculum Council (1990c) *Curriculum Guidance 6: Careers Education and Guidance.* York: National Curriculum Council.

National Curriculum Council (1990d) *Curriculum Guidance 7: Environmental Education.* York: National Curriculum Council.

National Curriculum Council (1990e) *Curriculum Guidance 8: Education for Citizenship.* York: National Curriculum Council.

National Curriculum Council (1991) *Geography Non-Statutory Guidance.* York: National Curriculum Council.

National Curriculum Council (1992) *Curriculum Guidance 9: The National Curriculum and Pupils with Severe Learning Difficulties and associated INSET resources.* York: National Curriculum Council.

National Curriculum Council (1993a) *An Introduction to Teaching Geography at Key Stages 1 and 2: NCC INSET Resources.* York: National Curriculum Council.

National Curriculum Council (1993b) *An Introduction to Teaching Geography at Key Stage 3: NCC INSET Resources.* York: National Curriculum Council.

National Curriculum Council (1993c) *Planning the National Curriculum at Key Stage 2.* York: National Curriculum Council.

OFSTED (1993a) *Geography Key Stages 1, 2 and 3, First Year, 1991-92.* London: HMSO.

OFSTED (1993b) *Geography Key Stages 1, 2 and 3, Second Year, 1992-93.* London: HMSO.

OFSTED (1994) *The Handbook for the Inspection of Schools.* London: HMSO.

Palmer, J. (1994) *Geography in the Early Years.* London: Routledge.

Peter, M. (1994) *Drama for All.* London: Fulton.

Rose, R. (1994) 'A modular approach to the curriculum for pupils with learning difficulties', *in*: R. Rose, A. Fergusson, C. Coles, R. Byers and D. Banes (eds.) *Implementing the Whole Curriculum for Pupils with Learning Difficulties.* London: Fulton.

Routh, C. and Rowe, A. (1993) *A Place for Stories at Key Stage 2.* Reading: Reading and Language Information Centre, University of Reading.

Rowe, A. and Routh, C. (1992) *A Place for Stories (Key Stage 1).* Reading: Reading and Language Information Centre, University of Reading.

School Examinations and Assessment Council (1993) *Children's Work Assessed: Geography and History.* London: SEAC.

Sebba, J. (1991) *Planning for Geography for Pupils with Learning Difficulties.* Sheffield: The Geographical Association.

Sebba, J. (1994) *History for All.* London: Fulton.

Sebba, J. and Clarke, J. (1993) 'Practical approaches to increasing access to geography', *Support for Learning,* 8, (2), 70-76.

Sebba, J., Byers, R. and Rose, R. (1993) *Redefining the Whole Curriculum for Pupils with Learning Difficulties.* London: Fulton.

Simkin, D. (1990) *Ourselves: An Infant Topic Pack for Integrated Humanities.* Brighton: Tressell.

Suffolk County Council (1991a) *Geography: developing mapwork skills at Key Stage 1 Guidance Booklet No 6.* Ipswich: Suffolk County Council Education Department.

Suffolk County Council (1991b) *Geography and Stories Guidance Booklet No. 3.* Ipswich: Suffolk County Council Education Department.

Suffolk County Council Humanities Advisory Team (1989) *Pontafon.* Ipswich: Suffolk County Council Education Department.

Waugh, D. and Bushell, T. (1992) *Key Geography.* Glasgow: Stanley Thornes.

Wiegand, P. (1992) *Places in the Primary School.* London: Falmer.

Wiegand, P. (1993) *Children and Primary Geography.* London: Cassell.

Wren Spinney School (undated) *Modules on 'Packaging and Advertising' and 'Learning about Maps'.* Kettering: Wren Spinney School.

Resources

ACTIONAID Education, 3 Church Street, Frome, Somerset BA11 1PW.

The Geographical Association, 343 Fulwood Road, Sheffield S10 3BP.

RNIB – Education Information Service, 224 Great Portland Street, London W1N 6AA.

Software suppliers

Northwest SEMERC, 1 Broadbent Road, Watersheddings, Oldham OL1 4HU.

Science Education Software, Unit 12, Marian Industrial Estate, Dolgellau, Gwynedd LL40 1UU.

Sherston Software, Swan Barton, Malmesbury, Wiltshire SN16 0LH.

Widgit Software, 102 Radford Road, Leamington Spa, Warwickshire CV31 1LF.